Metals and Ceramics

CONTEMPORARY
SCIENCE
PAPERBACKS
32

G. C. E. OLDS
B.SC., PH.D., C.ENG., M.I.MECH.E., F.I.M., F.INST.P.
Head of Metals and Ceramics Department
A.E.I. Central Research Laboratory

Metals and Ceramics

Dame Alice Owen's Girls' School

OLIVER & BOYD *Edinburgh*

OLIVER AND BOYD LTD
Tweeddale Court Edinburgh 1

First published 1968
© 1968 G. C. E. Olds

05 001745 4

Set in Times New Roman and printed in
Great Britain by Richard Clay (The Chaucer Press) Ltd
Bungay, Suffolk

Preface

The field covered by this book is intended to interest both general readers and those with some knowledge of physical science who are unfamiliar with recent research in metals and ceramics and its applications. The science of materials may be divided into the study of metals, ceramics and organic materials; metals and ceramics, despite superficial differences, have much in common in internal structure and properties, while the uses to which they are put are complementary and often similar.

In describing the essential character of metals and ceramics, it was thought desirable to define ceramic materials broadly, by reference to atomic bonding mechanisms, rather than be limited by a narrower definition related to a common method of preparing them – from powders by high temperature firing. Only the simple necessary concepts have been introduced in discussing atomic processes, with very brief reference to quantum considerations.

After describing various aspects of the behaviour of metals and ceramics, an account is given of some of the problems encountered and the results obtained in recent research and development on these materials. As the amount of both basic and applied research taking place is immense, subjects have been selected which are relevant to some major applications and which illustrate the use of new materials. An attempt has been made to be sympathetic to the reader in regard to scientific and technical terms peculiar to the subject.

I should like to make acknowledgement to colleagues, particularly those in my own Company, who have contributed to some of the items of research described, and also to those who have made helpful suggestions in reading through my text.

G. C. E. OLDS

Contents

Preface v
1. The Background 1
2. How Do Atoms of Metals and Ceramics Differ? 11
3. Metal Crystals and Ceramic Crystals 22
4. Microstructure – The Basis of Properties 33
5. Limitations of Metals and Ceramics 45
6. Strategy and Tactics of Materials Research 57
7. Metals in Gas Turbines and Steam Turbines 65
8. Materials in Nuclear Reactors 77
9. Ceramics in Electronics and Power Equipment 88
10. High Temperature and Space Technology 102
11. Metals and Ceramics Tomorrow 116
    Further Reading 129
    Index 130

1. The Background

Of all of the solid materials we use, if we exclude plastics and organic materials generally, the greater volume consists of just a few metals and ceramics – iron and steel, copper, aluminium, pottery and porcelain, bricks, tiles, stone and finally glass which I intend to classify with ceramics.

All of these materials, except aluminium, were known in some form in ancient times. But our ancestors, besides being unable to make high-quality materials in quantity, had no conception of the range of uses to which metals and ceramics could be put. Nor had they any idea of the newer metallic elements or ceramic compounds which are still used in relatively small quantity. The discovery and preparation of these had to await the growth of experimental science, the industrial revolution and the further technological revolution we are witnessing today.

The intriguing thing about 'materials science', to use modern terminology, is that the 92 chemical elements are basically all the materials we have to work with, if we ignore a few artificially made elements. Of the 92, 11 are gases and 3 are liquids at normal temperatures; a further 20 or so, some with picturesque names, are quite rare and still of specialised research interest only. Some 50 elements, mainly metals, and some of their simple compounds are the inorganic materials, the metals and ceramics, with which we shall mainly deal. But this large number of elements need not cause concern, since in Chapter 2 we shall classify them into a few groupings which will greatly simplify our understanding of them.

Before delving into the essential character of metals and ceramics, and the research on them which brings out their main properties and suggests uses, it is worth looking briefly at some recent history and established practice. The metallurgical and ceramics industries are concerned with obtaining high-quality minerals from the earth and using processes of treatment, refining and fabrication to give products such as steel or porcelain. Before the recent demands from the nuclear, electronics and aerospace fields for special and purer materials, almost all production was confined to a much shorter list of materials than we have now. Such common materials still form the bulk of production and themselves have been the subject of recent research.

The earliest metallurgy dated from the discovery of some native metals, but it was later found that metals could be produced from certain minerals when roasted or heated with charcoal (carbon). Iron oxide ore is reduced to the metal by coke in the modern blast furnace. The high-temperature reactions involve the coke, the oxygen in the air blast, and the removal of impurities by a liquid slag of mixed oxides. The cast 'pig iron' produced is impure and is remelted in a further refining process. Oxidation of impurities and reactions with the slag then occur to enable a 'cast iron' or a steel ingot to be made of controlled composition. The carbon content is very closely controlled in the case of a steel ingot, which may then be hot forged and perhaps rolled to give the product required.

The production processes for steel have advanced in recent years and some installations exist which eliminate the pig-iron stage, iron from the blast furnace being transferred directly to the refining furnace. Another advance, economically, has been to blow pure oxygen rather than air into the refining furnace for impurity removal. The higher-quality steel alloys are usually made by remelting in a 'basic electric' furnace with large additions of steel scrap of known quality. In this process, heating is by electric arcs struck between large carbon electrodes and the steel; the furnace is lined with 'basic refractories', e.g. magnesium oxide bricks. Temperatures of about 1600° C are reached in the molten steel.

A very recent innovation in steel production in Britain, necessitating only a fraction of the normal capital outlay, is 'spray refining'. Molten iron, direct from the blast furnace, is projected as a jet of particles through an oxygen atmosphere which purifies them before collection in a ladle and subsequently casting as a steel ingot.

Ordinary steel is an iron-carbon alloy which normally has impurities of manganese, silicon, sulphur and phosphorus. The addition of other alloying elements such as nickel, chromium or molybdenum has profound effects on its strength and toughness. A 13%-chromium addition renders it stainless. Much research has gone into steels of which there are hundreds of types to choose from today.

Some metallic copper occurs in nature, but much is now obtained from its sulphide ore which on heating yields the metal. Impurities are again oxidised away into a slag in a re-melting process, and some copper oxide which is formed is finally reduced back to the metal. High-purity copper is also made from impure copper by electrolytic deposition from solutions of copper salts. Pure copper is widely used as cable and wiring, its alloys with the metals cadmium, chromium or zirconium being used if high strength is required as well as good electrical conductivity. Alloys combining high strength with good corrosion resistance include cupro-nickel, the brasses (copper-zinc), copper-beryllium and the bronzes – originally copper-tin, but now including alloys containing aluminium silicon or lead. 'Copper' coinage is a copper-tin-zinc alloy, while 'silver' coinage is no longer a silver-copper alloy, but a cupro-nickel. 'Nickel silver' is a copper-nickel-zinc alloy, frequently electroplated with silver for ornamental ware (E.P.N.S.).

An important use for tin and zinc is the protection of steel sheet from rusting by surface tinning or galvanising, respectively. Other major outlets for tin are with lead as solder metal and in alloys for bearings. Zinc has uses for various domestic and industrial components, mainly as die-castings. The molten zinc alloy follows the fine detail of a metal die when cast under pressure; as zinc and its alloys have low melting points, the dies have a long life before wearing out. Lead has been widely used in bearing alloys, for pipes, roof protection, cable sheaths, batteries and for protection against atomic radiation.

Aluminium, magnesium and nickel have been widely used only in the present century, mainly due to the difficulties presented in their extraction and refining. There is much competition between the various metals for their major uses. Copper's original pride of place among the non-ferrous metals is being taken nowadays by aluminium, which has the advantages of lightness and cheapness as an electrical conductor. In alloy form it is a major material of construction.

Aluminium is mined as an impure oxide. This is purified and dissolved in a molten salt (sodium aluminium fluoride) which is then electrolysed at about 1000° C between carbon electrodes. The oxide is decomposed into oxygen and the pure metal, which collects at the bottom of the cell. Ingots are cast and sent to fabricators who roll or extrude the metal, or remelt and add other metals to obtain alloys. Aluminium is a fairly reactive metal, but its surface forms a very thin adherent oxide film which protects it from further corrosion. The metal forms strong light-weight alloys used as castings and wrought (mechanically fabricated) products. Its principal alloy additives are magnesium, manganese, silicon, copper and zinc.

In recent years non-electrolytic methods of extracting aluminium from its ore have been devised. Aluminium carbide or chloride is first formed and subsequently decomposed at high temperature.

Magnesium is another reactive metal which is also extracted by electrolysis from a fused salt mixture. One important source is sea-water, where considerable quantities are dissolved as the chloride. The metal is lighter still than aluminium, and its development and uses have accordingly increased with the growth of the aircraft and vehicle industries. It does not form a corrosion-resistant surface film as easily as aluminium, and is normally given a surface treatment for protection. In castings and wrought products, magnesium is frequently alloyed with small percentages of aluminium, zinc, manganese, zirconium or thorium.

Nickel extraction from its ores proceeds by separating an impure nickel sulphide from the other minerals present. The sulphide is roasted to give the oxide which is then reduced at

THE BACKGROUND

high temperature to the impure metal. The metal may be refined electrolytically, but in Britain the Mond process is used in which carbon monoxide first reacts with the nickel at about 60° C giving nickel carbonyl gas. This is then decomposed at about 180° C in a chamber containing nickel pellets on which the metal is deposited. The metal residues which do not combine with the carbon monoxide are rich as a source of the precious metals platinum and rhodium.

Nickel has properties similar to iron but is much more corrosion resistant, and its alloys compete with stainless steels in many applications. Its alloys with copper (cupro-nickels), often including other minor elements, provide a range of strong and corrosion-resistant materials, some of which are non-magnetic. These are used, sometimes as intricately-shaped components, by the food, chemical and many other industries. Nickel alloys containing substantial chromium and often iron contents are used for heating element wires and as materials with excellent strength and oxidation resistance at high temperatures. Nickel-iron alloys find use as low expansion alloys and magnetic alloys. Nickel is also added as a minor element in steel and cast irons, and forms a necessary underplating on which the more porous chromium plating is electrodeposited.

Practical metallurgy is much concerned with thermal and mechanical treatments to obtain metals in the required shape or form and with the necessary mechanical strength. Cast ingots of metals and alloys are often forged (pressed or hammered) by powerful machinery to a desired shape, and frequently rolled, either hot or cold, to form bar, plate, thin sheet or even foil. When originally cast, a metal is often inhomogeneous and brittle, but after it has been mechanically worked it has a more uniform and thus a stronger internal structure. The large-scale rolling of steel has been automated to give a controlled thickness to a few ten-thousandths of an inch. In the aluminium industry a modern process casts the metal in an endless length which is fed continuously from the casting machine to the rolling mill.

Many products such as containers or car bodies are formed

or fabricated from sheet steel by pressing with a hard steel tool. Other common components are castings of iron, steel, aluminium alloys and other non-ferrous metals. The highest-quality massive components are forged to nearly final dimensions and then precision machined. It is always a costly, but often necessary operation, to remove metal by machining, whether by traditional turning, milling, drilling and grinding, or by more modern electrochemical or spark-machining methods. Several new techniques of metal forming have been developed which promise to be important, especially when applied to metals which crack too readily in the course of ordinary pressing, rolling or extrusion. One such technique is 'high rate forming' which may utilise a fast-moving tool, or the shock from an explosive charge, to deform the metal to shape very quickly. Another is hydrostatic extrusion of bar or tube, which utilises a liquid at high pressure to force the metal through a die. High-frequency 'ultrasonic' vibrations can also be used to assist extrusion of brittle materials.

The joining of metals also has traditional and modern aspects. Apart from mechanical joints made by bolting or crimping by a pressure tool, metallurgical joining is done by soldering, brazing or welding. Common solder is a 60% tin-40% lead alloy, melting at about 185° C, and is used for simple electrical connections. For stronger joints which may also get hot during service, a braze metal such as copper-zinc-silver alloy melting at 700–800° C may be used. Welding is used when strong sound joints are required of similar quality to the parent metals being welded. This involves locally melting the surfaces to be joined and often introducing a filler metal similar to the parent metal. An oxygen-enriched gas flame may be used for welding steel at about 1500° C, but it is common to employ an electric arc of several hundred amps struck between a welding electrode and the surface to be welded; a consumable electrode of filler metal, coated by a suitable flux to form a protective slag on the molten metal, is often used. A tungsten electrode may also be used, a separate rod of filler metal being fed into the arc if required, and the whole enclosed by an argon inert gas shield. Argon-arc welding is commonly used for aluminium joints which have to bear stress during service. Another process is 'resistance welding' which makes use of the heating effect of an electric current between the metal parts to be joined.

Many new and specialist welding techniques are now available. Electroslag welding (Fig. 1) utilises heat produced by a large current passing through the slag between the workpiece and a consumable electrode. This is used for large welds in thick steel sections. A large-scale electroslag process has also recently been developed for refining steel: the metal passes from the electrode to the deposit, where it builds up as a

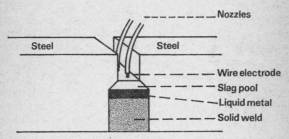

Fig. 1. *Electroslag welding. The wires, fed through the nozzles, melt in the slag pool. The weld is cooled by water flowing through copper shoes at front and back (not shown).*

homogeneous high-quality casting. In pressure welding, the two components are deformed so that the mating surfaces expose clean new areas in intimate contact which then cohere. A similar process occurs in explosive welding. In friction welding, relative movement or rotation generates heat locally to melt a very thin layer on the two surfaces to be joined. Other methods of melting a thin layer of the metals to be joined are electron beam and laser welding, which use directed narrow beams of charged particles and optical radiation respectively.

A totally different approach to making components, especially suitable for mass production of small metal shapes, is the technique of powder metallurgy. Iron powder, for example, can be pressed to the required shape and 'sintered' in a furnace, well below its melting point, so that it shrinks to form a homogeneous solid, often of adequate dimensional accuracy for direct use as a component. Certain high melting-point metals and many ceramics are most conveniently made or consolidated by some powder technique followed by high-temperature sintering. Still different techniques are used to prepare pure forms of other less common metals on which much research is now taking place.

The field of ceramics and glass is concerned to a large extent with the oxides of the chemical elements. A unique glass, commonly called fused quartz, consists entirely of the oxide of silicon (silica) and can be made by melting quartz or sand at over 2000° C. But the use of such a high temperature results in a relatively costly product. More usual glasses have a high silica content, but much lower melting and softening temperatures. This can be achieved by adding an alkali flux, such as soda, to the sand (silica), so that melting and refining (mainly bubble removal) can take place at about 1400° C. The common window and bottle glass of today has a soda-lime-silica composition, over 70% by weight being silica. Glass articles are sometimes cast, or may be fabricated at a few hundred degrees below the melting temperature by pressing, blowing or shaping the viscous mixture in some way.

Naturally occurring minerals formed the basis of the earliest glass and ceramics made. The fluxing of the common oxide mineral, silica, with other oxides to form a glass is a rather special occurrence in that most oxides and other ceramic compounds do not form glasses. Instead, they often have very high melting points and are now normally consolidated into shapes by powder compaction followed by sintering. Nevertheless, many common ceramic articles, such as porcelain, are a mixture of oxides, including silica, and do contain some glass.

Porcelain is made by firing a mixture of clay, sand (silica) and an alkali flux such as feldspar. The latter is a common major constituent of stone or rock, consisting of soda or potash (sodium and potassium oxides), alumina (aluminium oxide) and silica. Clay consists essentially of alumina and silica and results from the weathering of rock, during which the latter loses its alkali content. A mixture of clay and water has unique plastic properties and, when mixing with the other powdered material of suitable particle size, can be shaped and ultimately fired in a kiln. At the sintering temperature, the alkali flux gives rise to a molten glass which partly dissolves the other oxide constituents and binds them together on cooling (see Fig. 2). Such articles can also be glazed by applying to the surface an appropriate powder mix which forms a glass on firing. Colours and stains may be incorporated in the glaze. Enamelling is a similar process often applied to metal surfaces, and normally giving a mixed glass and ceramic coating.

Fig. 2. *Porcelain under microscope* (× *1000*). *The background is an alkali/alumina/silica glass which binds together large grains of quartz* (white) *and small needle-like grains of mullite* (*an alumina/silica compound*). *The black areas are pores which did not escape during firing.*

Glass and common pottery are not strong mechanically, but quite strong porcelain compositions are made, for example, by adding extra alumina or some zirconia (zirconium oxide). Good electrical insulation properties, as well as adequate strength can be obtained if magnesia (magnesium oxide) is combined with alumina and silica in the absence of alkali.

Bricks, tiles and ceramic pipes are relatively cheap products made from the less pure heavy clays. A suitable mix is cast to shape, dried and fired. The products are somewhat porous, but a glaze can be applied if required. Portland cements, which are lime-silica-alumina materials, acquire strength at room temperature by slowly absorbing water which leads to a chemical change giving a compact solid material. Subsequently heating such cement to a few hundred degress would dehydrate it and its strength would be lost.

Another traditional class of ceramics are known as refractories. These are bricks and other structural members used in high-temperature furnaces. Typical requirements for refractories are high strength and resistance to corrosion and erosion by molten glass, metals, slags and high-temperature gases. Bricks based on high melting-point oxides are frequently used and, dependent on whether acidic or basic refractories are required, may utilise silica or magnesia respectively. The latter are more important for the higher furnace temperatures now being used in the steel industry. Recent research has shown that bricks with a very high magnesia content, sometimes with a chromic oxide addition, give much improved strength at high temperature.

More expensive refractories may utilise alumina or zirconia, while for the highest temperatures there is thoria (thorium oxide) although it is little used, which melts at 3300° C. Oxide refractories, however, are not always the best high-temperature materials in particular cases, for example when the atmosphere is non-oxidising or when sudden temperature changes may produce cracking due to 'thermal shock'. An important non-oxide refractory ceramic is silicon carbide, or carborundum, which can be used in copper-melting furnaces up to 1200° C.

Many of the ceramics of advanced technological interest, on which much research has been taking place, are in a different category to the more traditional ceramics so far described. The newer materials are often of high purity or closely controlled composition in which great care is taken to obtain special properties such as strength or certain electrical behaviour. Oxide ceramics are perhaps of greatest interest, but other important ceramics are compounds of the elements with carbon, silicon, nitrogen and boron, as well as the elements carbon and silicon themselves. It is the character and uses of the newer class of ceramics that are of great research interest today.

2. How Do Atoms of Metals and Ceramics Differ?

If a typical piece of metal or ceramic is examined, distinct sets of properties could be drawn up (Table 1) which appear to differentiate the two clearly. The most characteristic feature is that a metal is ductile, while a ceramic is brittle. Yet even this most simple generalisation is not always true, and there are some metals and ceramics which far from conform to all the properties listed. To appreciate the essential differences and to understand recent research and its applications, it is necessary to examine some of the most basic properties of the chemical elements. If a little time is taken at this stage to study the main groups into which the 92 elements are classified, the subject will appear both more simple and more intriguing.

Long before many of the elements were discovered the hypothesis was put forward that if a piece of matter were continually subdivided a very small particle of matter would ultimately be reached which could not be further subdivided without altering its basic nature and properties. Such a particle is an atom, and its very small diameter is now known to be between about 2 and 5 Å (1 Å, or Ångstrom unit being 1/100 000 000 of a centimetre, i.e. 10^{-8} cm). All matter occurring in nature consists of atoms of one or more of the 92 elements. Each has been given a symbol, e.g. Al for an aluminium atom.

For some purposes the atoms may be regarded as minute 'billiard-ball'-like objects, but in fact each consists of a central 'heavy' nucleus surrounded by a number of orbiting electrons, which we may consider as very small particles. The 92 elements in succession have from 1 to 92 electrons per atom. Each has a number of electrons characteristic of that element, for instance hydrogen has one electron and iron has 26. The 'atomic number' of the element derives from this. The electrons each

Table 1. *Typical properties of metals and ceramics.*

Metals (e.g. Copper)	Ceramics (e.g. Alumina)
Soft and non-abrasive	Hard and abrasive
Ductile and tough	Brittle
High tensile and compressive strength	Weak in tension but with high compressive strength
Electrical conductor	Electrical insulator
Good thermal conductor	Thermal insulator
Withstands thermal shocks	Cracks readily in thermal shock
Softens at high temperature	Refractory at high temperature
Tarnishes and is chemically reactive	Stable and inert to environment
Homogeneous and rings when struck	Non-homogeneous and absorbs sound

have a fixed negative electrical charge, while the larger nucleus of the atom has a positive charge equal to the total charge on the electrons which orbit around it. So the net total electrical charge on each atom is zero.

Table 2 is a form of the 'periodic table' of elements, giving atomic number and symbol in each case. This long list, for completeness, contains all the natural elements, but it is immediately simplified by classifying similar elements into groups – eight A groups and eight B groups. For instance, Group VIIIA contains the inert gases, all with similar properties but quite different from elements in the other groups.

Before examining the periodic table further, some fundamental laws of physics and chemistry must be mentioned. The first is that the total energy possessed by atoms, including energy due to electrical charges always becomes the minimum value possible.* The law incidently applies also to the gravitation energy of a piece of material suspended above ground level; if permitted, it will fall to the ground where it has lower energy, having lost the energy it had due to height. Similarly, the 'electrostatic' attraction between a positively

---

* Strictly, nature is not quite so orderly as this, and the law should contain a factor for disorder of atoms due to temperature, as mentioned on page 30.

HOW DO ATOMS OF METALS AND CERAMICS DIFFER?

charged particle and a negatively charged particle may be seen as due to the minimum energy law. Their energy is reduced by being attracted together, into contact if possible. It therefore may appear surprising that the negatively charged electrons orbiting the nucleus of the atom are not attracted by the positive nucleus sufficiently for them to fall into it. In fact, it was found earlier in this century that the classical laws of physics were inadequate to deal with the ultimate particles of matter, and that different 'quantum mechanical' laws must be applied. According to these, each of the atomic electrons must retain a definite amount of energy as it orbits the nucleus. As electrons are added to the atom, to give elements of higher atomic number, they may be considered to form a series of shells around the nucleus, each shell containing a fixed maximum number of electrons, in many cases, eight. Thus the electrons in an atom will always reduce their energy by 'falling' towards the nucleus, into the innermost shell that the quantum mechanical laws allow. It is the electrons in the outer shell of an atom which largely dictate the chemical properties of the element.

Referring to the periodic table, hydrogen (H) is the simplest atom, having one electron, while helium (He) has two electrons. At this point this first electron shell, which is nearest of all to the nucleus and can contain a maximum of two electrons, is full and further electrons must go into the second shell. The second shell can contain up to eight electrons and gives the second 'period', of eight elements. Lithium (Li) has its inner shell full like helium and one extra electron in its outer shell (see Fig. 3). Going along the second period in Table 2, beryllium (Be) has two electrons in this shell, boron (B) three and so on until neon (Ne) is reached with eight, which fills the shell. The third period from sodium (Na) to argon (Ar) fills the next shell which also can contain up to eight electrons. Looking at the table in columns, the elements mentioned so far are in the eight A groups. The next three periods of the table have 18 elements each, due in each case to an extra sub-shell which contains up to ten electrons, giving rise to the B groups. One irregularity here is that 14 extra 'rare-earth' elements, of

Table 2. *Periodic table of elements.*

PERIOD \ GROUP NUMBER	IA	IIA	IIIB	IVB	VB	VIB	VIIB	VIIIB			IB	IIB	IIIA	IVA	VA	VIA	VIIA	VIIIA
First Shell or Period	1 H																	2 He
Second Shell or Period	3 Li	4 Be											5 B	6 C	7 N	8 O	9 F	10 Ne
Third Shell or Period	11 Na	12 Mg											13 Al	14 Si	15 P	16 S	17 Cl	18 Ar
Fourth Period	19 K	20 Ca	21 Sc	22 Ti	23 V	24 Cr	25 Mn	26 Fe	27 Co	28 Ni	29 Cu	30 Zn	31 Ga	32 Ge	33 As	34 Se	35 Br	36 Kr
Fifth Period	37 Rb	38 Sr	39 Y	40 Zr	41 Nb	42 Mo	43 Tc	44 Ru	45 Rh	46 Pd	47 Ag	48 Cd	49 In	50 Sn	51 Sb	52 Te	53 I	54 Xe
Sixth Period	55 Cs	56 Ba	57 La	72 Hf	73 Ta	74 W	75 Re	76 Os	77 Ir	78 Pt	79 Au	80 Hg	81 Tl	82 Pb	83 Bi	84 Po	85 At	86 Rn
Seventh Period	87 Fr	88 Ra	89 Ac	90 Th	91 Pa	92 U												

| The Rare Earth Elements | 58 Ce | 59 Pr | 60 Nd | 61 Pm | 62 Sm | 63 Eu | 64 Gd | 65 Tb | 66 Dy | 67 Ho | 68 Er | 69 Tm | 70 Yb | 71 Lu |

HOW DO ATOMS OF METALS AND CERAMICS DIFFER?

atomic numbers between 58 and 71, appear in the sixth period, due to a further sub-shell which can contain 14 electrons. Atomic weight goes up increasingly with atomic number, uranium being the heaviest atom in the table, with 238 times hydrogen's atomic weight.

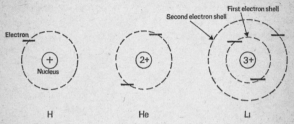

Fig. 3. *Structure of atoms. Diagrammatic representation of atoms of hydrogen, helium and lithium.*

A dozen or so artificial elements have also been prepared, taking atomic numbers to well over 100. These are all unstable, only plutonium (Pu) of atomic number 94 being of major interest.

If we ignore the B-group elements for the moment, and consider the main eight-electron shells of the A groups, all Group IA elements have one electron in their outermost shell, all Group IIA elements two, and so on up to Group VIIIA elements, which all have full shells. The outer electrons in partly filled shells are called 'valence' electrons and govern the chemical reactivity of the element. Thus elements in the same group tend to have quite similar properties.

The valence electrons represent a definite increase in energy of the atom above that due to its already full shells, so according to the minimum energy law, atoms of low valency try to give up these valence electrons to other atoms requiring a small number of electrons to fill their shells. For instance, the sodium (Na) atom in Group IA will readily give up its single valence electron to a chlorine (Cl) atom in Group VIIA so that both have full outer shells with a net reduction in energy. If this occurs, the sodium atom has lost, and the chlorine atom has gained, the negative charge of the transferred electron. The total electric charge of the orbiting electrons for each atom now no longer exactly balances the positive charge on the nucleus. Such an atom with a net electric charge is called an ion; in this case the sodium atom becomes a positive sodium ion, written Na^+, and the chlorine atom becomes a negative chlorine ion, Cl^-. As each have full shells, such ions are unreactive chemically, like the inert gases of Group VIIIA, but as they are charged electrically the positive and negative ions attract each other strongly and form the chemical compound sodium chloride NaCl. This is common salt, which is quite stable, although its constituents (sodium and chlorine), due to their original odd electron either in excess or wanted, are very reactive and are biological health hazards.

In a similar manner magnesium (Mg) in Group IIA will readily lose its two valence electrons to oxygen (O) in Group VIA, which requires two to complete its shell. This chemical reaction is quite a spectacular example of the minimum energy law. It occurs when a piece of magnesium foil burns in oxygen or air at white heat forming the compound magnesium oxide or magnesia (MgO), consisting of the ions Mg^{2+} and O^{2-}. Magnesium may equally give its two valence electrons to two chlorine atoms, each requiring one, to form the stable compound $MgCl_2$. Another example of spare and wanted electrons balancing out to form a stable compound is aluminium oxide, or alumina (Al_2O_3), where two Al^{3+} and three O^{2-} ions combine.

Moving across the periodic table from left to right, the atoms have more valence electrons and are said to become more electronegative in character. For other reasons there is also a slight increase in electronegativity for many groups when moving from the bottom to the top of the table. A little further study of the elements will connect the nature of metals with low valency and least electronegativity; the more electronegative elements are in the upper right-hand corner of the periodic table and are non-metals.

The electron shell structure in the B-group elements is a little more complex than in the A groups. Group IB, containing copper (Cu), silver (Ag), and gold (Au), and Group IIB

HOW DO ATOMS OF METALS AND CERAMICS DIFFER? 17

(zinc, etc.) bear some similarity in electron structure to Groups IA and IIA, and their valencies are frequently one and two respectively. Group IIIB elements have a valency of three and form stable oxides such as yttria (Y_2O_3) and lanthana (La_2O_3). Elements in the higher B groups have various valencies, but not usually of high value; for example, iron, in Group VIIIB may have divalent (Fe^{2+}) or trivalent (Fe^{3+}) ions. Groups IVB, VB and VIB contain elements of some special interest which also form important refractory compounds with elements in the A groups, e.g. zirconium oxide, or zirconia (ZrO_2), and tungsten carbide (WC). Group VIIIB contains the three quite similar magnetic elements, iron (Fe), cobalt (Co) and nickel (Ni), and also the six precious 'platinum metals'.

The basic difference between metals and ceramics is closely connected with the role of electrons in binding the sub-microscopic atoms together to form the macroscopic visible solids we can handle. The outer semi-filled shell of valence electrons leads to three major types of interatomic bond; these are called ionic, covalent and metallic or free electron bonds.

The ionic bond occurs in the case of the compound sodium chloride (NaCl), mentioned previously; here two different atoms become oppositely charged ions which are bonded together by electrostatic attraction (see Fig. 4). The cohesion between alternate sodium and chlorine ions is repeated in three dimensions to give a large grain of sodium chloride. Divalent compounds such as magnesia (MgO) are ionically bonded by two electrons which form a stronger bond, giving greater stability and a higher melting point than occurs in univalent compounds like NaCl. Alumina (Al_2O_3) is a refractory compound with bonds between Al^{3+} and O^{2-} ions. Ionic bonds are quite strong and form between atoms of relatively low and high valency respectively. All such ionically bonded compounds are classified as ceramics.

A second way the electrons may serve to bond the atoms of a solid together is by covalent bonding. When this is strong, stable high melting-point elements or compounds occur which we shall also classify as ceramics. Covalent bonding occurs when electrons are shared between the outer shells of neighbouring atoms; the bond results from the minimising of total energy thereby achieved. It exists between atoms of the same or nearly the same valency, widely differing valencies usually giving the ionic bond. Good examples of covalent bonding occur in Group IVA in diamond (one of the pure forms of carbon) and silicon (Si). Here the four valence electrons of each atom are shared with the four atoms immediately surrounding it. Thus each valence electron may be regarded as shared between two neighbouring atoms, so that each atom has a filled outer shell of eight shared electrons. The electrons form the bond by orbiting in a shared outer shell around their local positive 'ion' cores which mutually attract them. Silicon carbide (SiC) is a covalent compound, again with each atom providing four electrons. Pairs of elements with valencies of three and five can also form strongly covalent compounds such as boron nitride (BN) or aluminium nitride (AlN). Such materials are also ceramics. A degree of covalency also occurs in many ionically bonded ceramics, e.g. oxides.

Covalent bonding also exists for a whole range of 'organic' compounds largely based on carbon, which are not considered in this book; such compounds, when they are solids, are generally less strongly bonded and less stable at elevated temperature than the covalent ceramics.

The metallic bond occurs most characteristically between atoms of low valency when there are not enough electrons to form local covalent bonds. Instead, the valence electrons are shared by all the atoms in the solid. They become free of their particular atoms and may be regarded as in a broad unfilled shell surrounding all the positive ions and holding them together by attraction. This type of bonding applies to both elements and their alloys, and does not obey any valency rules. Certain alloys, however, may contain 'intermetallic compounds' of elements which have some degree of covalent, as well as metallic bonding, for example the magnesium-tin compound, Mg_2Sn.

Ignoring hydrogen, which is a special case, all the elements of Groups IA and IIA are bonded by free electrons; these are

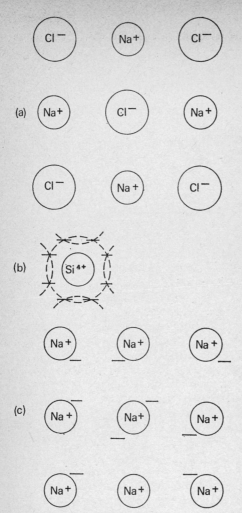

Fig. 4. *Types of bond between atoms.* (a) *Ionic bond in sodium chloride – electrostatic attraction between Na^+ and Cl^- ions.* (b) *Covalent electron-sharing bond in silicon between the Si^{4+} 'ion' and its four neighbours (shown in one plane in diagram).* (c) *Free-electron bond in metallic sodium.*

the so-called alkali and alkaline-earth metals, respectively. All the B-group elements are also metals, but as we proceed across the periodic table from Group IIIA, covalent bonding takes over from free electron or metallic bonding. We shall regard as ceramics the solid elements in which the atoms are joined by fairly strong covalent bonds. The transition from metallic to covalent bonding is not sharp, and there are elements with both bonding mechanisms. At the right-hand side of the periodic table, solid elements exist in which the bonding is partly covalent and partly by a weak mechanism called the van der Waals bond. The latter is due to electrons in the outer shells of neighbouring atoms moving in sympathy with each other; this results in a slightly lower total energy, and therefore a force of attraction between the atoms. Such elements, e.g. iodine, may be regarded as organic solids. Whether we regard an element as metallic, ceramic or organic ultimately depends on our definition of these terms, since some elements and their compounds do not fall into clear categories.

However, we shall consider all inorganic solids as either metals or ceramics. Metals may be defined as solids which are free electron-bonded with little or no other bonding mechanism present. Ceramics, then, are all solids in which the bonding mechanism consists of significantly strong ionic or covalent bonds, even though in some cases other bonding mechanisms are also present.

In Groups IIIA and IVA, both metal and ceramic elements occur. In Group IIIA boron (B) is covalently bonded and is a ceramic, while aluminium (Al) and the rest of the group are metals. In Group IVA carbon (C), silicon (Si) and germanium (Ge) are covalently bonded and, within our definition, are ceramics, while the normal form of tin (Sn) and lead (Pb) are metals. Moving to the right of the periodic table, the solid elements of Groups VA, VIA and VIIA are largely covalently bonded, the van der Waals bond playing an increasing role, and also occurring when the gases nitrogen (N), oxygen (O), fluorine (F), chlorine (Cl) and the liquid bromine (Br) are frozen. In Group VIIIA the inert gases, when frozen at very low temperature, are bonded solely by the weak van der Waals

HOW DO ATOMS OF METALS AND CERAMICS DIFFER?

bond. However, in Group VA, the covalent bond becomes more metallic in character from arsenic to antimony and bismuth; this applies to a lesser extent in Group VIA for selenium (Se), tellurium (Te) and polonium (Po). Although some may refer to the more metallic of these partly covalent elements as metals (e.g. arsenic, antimony and bismuth), our definition classifies them as ceramics, while the least strongly bonded solids of Groups VA, VIA and VIIA (e.g. phosphorus, sulphur and iodine) are regarded as organics.

Thus the nature of the bonds which particular atoms form with each other, or with atoms of other elements to form compounds, answers in the simplest way the 'loaded' question which forms this chapter heading. Atoms of metals in Groups IA, IIA, IIIA or the B groups form ionic bonds with atoms of Groups VIA and VIIA giving ceramics such as NaCl or MgO; atoms of Groups IVB, VB and VIB metals also form bonds with the atoms of Groups IIIA, IVA and VA giving ceramics, such as titanium nitride (TiN), molybdenum boride (MoB_2) or niobium carbide (NbC), where the bond is partly covalent and partly metallic, sometimes with slight ionic bonding. Atoms of metals will, however, mix with each other to form free electron-bonded alloys over wide ranges of composition, not governed at all by the valency rules.

Reference again to Table 1 will remind us how the elements, alloys and compounds defined as metals and ceramics conform to their supposed typical properties. In particular, all metals and alloys have some degree of softness, or ductility, and their free electrons give rise to both good electrical conductivity and chemical reactivity towards other elements, e.g. oxygen of the air. Ceramics, on the other hand, are hard and brittle materials which are often chemically inert and poor electrical conductors; these two properties result from their valence electrons being used, to varying extents, to form ionic and covalent bonds.

3. Metal Crystals and Ceramic Crystals

Having discussed something of the internal structure of atoms and how this leads to different types of bond with other atoms, giving us a metal or a ceramic, it is now possible to look at the ways in which large numbers of atoms join together to form the macroscopic solids we handle. For this purpose atoms and ions of the elements may be regarded as minute 'billiard-ball' spheres of various sizes. The ions, having gained or lost electrons, are larger or smaller spheres respectively than their parent atoms.

Unlike the atoms of a gas or liquid, atoms of a solid normally take up orderly positions in straight rows extending into three dimensions. When, for example, a molten metal of a single element solidifies, atoms (or strictly, ions) of the same size pack together to form a regular array giving their most stable (minimum energy) configuration. It is such assemblies of atoms that are called crystals; all solid metals and ceramics, except glass, have such a crystalline structure.

There are three principal ways in which atoms of metals pack together in rows. The two most closely packed structures are the face-centred cubic (FCC) and close-packed hexagonal (CPH) crystals. The third common crystal structure, which is not quite so close-packed but is favoured by some metals, is called body-centred cubic (BCC). The origin of these terms is clear from the diagrams of 'unit cell' patterns where the atoms are drawn as relatively small circles (Fig. 5). The structure of the unit cell is repeated throughout the crystal.

The particular crystal form which the different metals take depends on details of electron structure. Metals which form

BCC crystals include all the Group IA alkali metals, iron and six 'high-temperature metals' of Groups VB and VIB: vanadium (V), niobium (Nb), tantalum (Ta), chromium (Cr), molybdenum (Mo) and tungsten (W). In BCC crystals, each atom has eight other atoms all equidistant from it; perhaps we can see overtones of the covalent bond in the univalent alkali metal atoms having eight nearest neighbours. Atoms of the FCC and CPH structures have 12 nearest neighbours, which

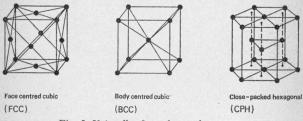

Face centred cubic (FCC) Body centred cubic (BCC) Close-packed hexagonal (CPH)

Fig. 5. *Unit cells of metal crystals.*

is the number expected for closest packing of spheres when there is no valency restriction. FCC metals include the Group IB metals (copper, silver and gold) and aluminium, nickel, lead and a high temperature form of iron. Common CPH metals, including some hexagonal metals which are not quite close-packed because the relative height of the hexagon is slightly different, are beryllium, magnesium, zinc, cobalt, and the Group IVB metals titanium, zirconium and hafnium. We shall see later that these three simple structures lead to ductility in metals.

Turning to ceramics and first to the strongly covalent Group IVA elements, carbon (in the form of diamond), silicon and germanium form the so-called diamond (a modified FCC) structure in which each atom has four nearest neighbours. Silicon carbide (SiC) can form both this structure and a hexagonal structure. The more common form of carbon (graphite) and that of boron is hexagonal.

In ceramic compounds which are ionic we find that the positive ion (or cation) of the metal is usually smaller than the negative ion (or anion) of the non-metal such as oxygen or chlorine. The crystal structures which are formed by many ceramics are based on those already mentioned for metals, but the relative location of anions and smaller cations depends on several factors: the different ion sizes, any degree of covalency present and the need to balance out the positive and negative ionic charge locally.

There are two important types of ceramic crystal in which the anion forms an FCC structure and the metal cations lodge

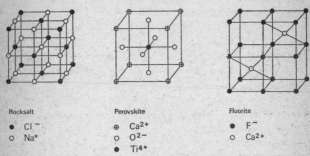

Rocksalt
- Cl⁻
○ Na⁺

Perovskite
⊕ Ca^{2+}
○ O^{2-}
● Ti^{4+}

Fluorite
● F⁻
○ Ca^{2+}

Fig. 6. *Some ceramic crystal structures.*

in certain of the interstices between the anions: in the first, the *rocksalt* or NaCl structure, the addition of sodium to the chlorine FCC lattice results in all the ions present forming a simple cubic pattern (Fig. 6); many of the alkaline earth oxides (MgO, CaO, SrO, BaO) as well as MnO, FeO, CoO and NiO have this structure. In the second, the structure named after the double oxide mineral *spinel* ($MgO.Al_2O_3$), the two different cations go into different sets of interstices in the oxygen FCC structure. The most important magnetic ceramics have this type of structure.

The mineral *perovskite*, or calcium titanate ($CaO.TiO_2$, or $CaTiO_3$), forms a structure shared by other ceramics which have special electrical properties, e.g. strontium, barium and lead titanates ($SrTiO_3$, $BaTiO_3$, and $PbTiO_3$). In perovskite the oxygen anions and calcium cations form an FCC structure with the small titanium ion fitting into the cube centre.

Two other important crystal forms have an anion CPH structure; one is the *corundum*, or alumina (Al_2O_3) structure which is shared by chromic oxide (Cr_2O_3), ferric oxide (Fe_2O_3) and magnesium titanate ($MgTiO_3$); the other is that of the mineral *wurtzite*, a form of zinc suphide (ZnS), which is also the structure of beryllia (BeO) and aluminium nitride (AlN). In corundum the metal cations are so positioned that they have six nearest neighbour anions, while in wurtzite they have four. Finally, the mineral *fluorite*, or calcium fluoride (CaF_2), gives its name to a structure shared by the refractory oxides, urania (UO_2), thoria (ThO_2) and an important form of zirconia (ZrO_2). Here the anion forms a simple cubic structure with a cation in the centre of alternate cubes.

It is easy to see how a metal becomes alloyed when atoms of the host metal are replaced in the crystal 'lattice' by foreign atoms of similar size. If 30% (by weight) of zinc is dissolved in molten copper and the resulting alloy is solidified, zinc atoms replace many copper atoms in the FCC lattice, giving a 'solid solution' of zinc in copper (common brass). But if 40% is added, the strain on the FCC lattice due to the larger zinc atom, together with other effects due to valence electrons, leads to the formation of another crystal structure in the solid solution. This structure is BCC and consists essentially of the intermetallic compound CuZn. Thus a 60% Cu-40% Zn brass consists of two 'phases', the BCC compound CuZn and the FCC copper-rich phase containing the maximum amount of zinc (about 35%) it can tolerate in solid solution. There is similarly a limit to the solid solubility of carbon in iron, but in this material (steel) the small carbon atoms dissolve interstitially in the lattice instead of substituting for iron atoms.

Ceramics do not form solid solutions so readily as metals since the ionic valencies, besides ionic sizes, must be compatible. In other words, the added ions have to possess the 'correct' electrical charge to preserve electrical neutrality locally if they are to take up substitutional positions in their host crystal. Thus magnesia (MgO) will dissolve to some extent in a similar divalent oxide, calcia (CaO), if the two powders are sintered at high temperature. But the addition of magnesia to alumina (Al_2O_3) immediately leads to a small amount of the spinel phase, of different crystal structure to the matrix.

So far, crystals of metals and ceramics have been considered to be perfectly ordered lattices. In fact, some of their most important properties arise as a result of the many imperfections they contain (Fig. 7). A metal may contain what are

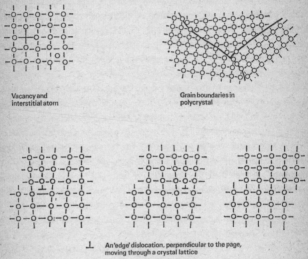

Fig. 7. *Imperfections in crystalline structure.*

called point defects, line defects and surface defects. A point defect may be a 'vacancy' or absence of an atom from its correct site, or it could be an interstitial atom, perhaps of a 'foreign' element, which has become lodged at the 'wrong' point in a crystal. In either case a point defect distorts or strains the lattice in its vicinity and thus introduces extra energy; such energy may be provided by heat and, at a given temperature, there may be one vacancy in many millions of atomic positions, i.e. perhaps 10^{16} vacancies per cm^3. Line defects or 'dislocations' occur when rows of missing atoms occur giving rise to a linear fault. Small movement of atoms in the strained region of the fault can allow the dislocation to move through the crystal. In doing this, the dislocaton has left a perfect crystal behind it, but the whole crystal has 'slipped' by one atomic spacing along one of its planes. Such slip can easily be produced by dislocations on many crystal planes in the simple crystal structure of metals; it is this which gives metals their basic ductility or plasticity.

The other important defects consist of internal surfaces. When a metal freezes or solidifies from the melt, it does not form only one single crystal unless very special care is taken; instead, many small crystals form, or 'nucleate', throughout the liquid, and the atoms attach themselves to one or other of the growing crystals until the whole mass is solid. This produces a 'polycrystalline' solid consisting entirely of crystals or grains with random angular orientations and shapes. The size of the grains in a given piece of metal is normally fairly uniform, a typical mean diameter being 10 or 20 microns (μ) ($1\mu = 10^{-4}$ cm). Grain boundary surfaces form a major defect between adjacent crystal lattices and, like dislocations, affect the mechanical behaviour of the metal.

Similar defects are found in ceramics, although it is noteworthy that nature provides some large single crystals, e.g. rocksalt (NaCl), diamond (C) and sapphire (Al_2O_3). When compared with a metal, point and line defects have certain constraints in a normal ceramic polycrystal. If, for example, we produce some oxygen vacancies in uranium dioxide (UO_2) by heating it in an oxygen-free atmosphere, its composition might become $UO_{1.99}$ which is not a 'stoichiometric' compound, i.e. obeying the valency rules (U^{4+} and $2O^{2-}$). So some uranium atoms present, to preserve electrical charge neutrality, would become lower valency ions. Even this would not be a very stable low-energy arrangement, and it is common in ceramics for vacancies to occur at adjacent cation and anion sites to maintain local charge neutrality. A similar problem occurs if a foreign ion of the 'wrong' valency is substituted in the crystal lattice.

The presence and movement of dislocations is also made difficult in ceramics by their more complex crystal structure and by the need for local charge neutrality between adjacent layers of ions. An exception is found in certain single crystals such as sodium chloride and magnesium oxide which have the simple rocksalt structure in which slip will occur on certain crystal planes. But many other single crystal ceramics and all normal polycrystalline ceramics do not exhibit slip and are brittle. The energy relationships are such that the atomic planes would sooner physically separate by fracture, than slip.

In a metal crystal there are sets of planes at various angles along which dislocations can move and produce large amounts of slip. When a dislocation is obstructed by a grain boundary in a polycrystalline metal, several more dislocations may pile up behind it at the end of the slip plane. The local strain energy then becomes so great that dislocations are generated in the next grain and so on. All this happens almost instantaneously as a piece of metal is deformed or strained plastically. As this occurs, the dislocations on the various planes of a crystal get tied up with each other and the crystals become distorted. A higher stress then has to be applied to strain the polycrystal further, and the metal is said to have 'strain-hardened' or 'work-hardened'. At a still higher stress, it is energetically easier for the metal to fracture than deform further. This stress is the fracture strength of the metal, while the stress at which it first deforms plastically is called the yield strength. Typical yield and tensile fracture strength figures for normal polycrystalline copper are 20 000 and 35 000 psi (i.e. pounds per square inch). For many steels these figures are over 100 000 psi while the tensile strength of a high-quality alumina is 40 000 psi; alumina, being brittle, does not have a yield strength.

When the stress applied to a solid is too low to cause plastic (permanent) deformation, it always undergoes a small 'elastic' deformation or strain which disappears when the stress is removed. This is due simply to compression or dilation of the crystal lattice; the atoms may be regarded as slightly soft and elastic. The amount of tensile elastic strain (percentage increase in length) a solid can tolerate before yield or fracture varies, but it is usually less than 1%. Such small elastic strains are always proportional to the stress applied, and the ratio of tensile stress to elastic strain is called Young's modulus of elasticity; this is different for different materials, and its magnitude is about 30 000 000 psi for all steels. Thus an elastic strain of 0·1 % in a bar of steel would require a stress of 0·1% of 30 million psi, i.e. 30 000 psi.

Another basic property of matter concerns its temperature. At the 'absolute zero' of temperature ($-273°$ C), all atoms are effectively at rest, but as heat is given to a solid its atoms absorb this as 'kinetic energy' of agitation. In a crystal the atoms vibrate elastically about their lattice positions and it is this 'thermal vibration' which we interpret as the solid's temperature. The energy of movement of an individual atom may vary somewhat from the mean, and when the temperature is high enough the more energetic ones can jump into a nearby vacancy or even into an interstitial site in the crystal. There are always a number of vacancies due to thermal agitation of atoms in a crystalline solid, and their number increases with temperature. The random jumping of atoms into nearby vacancies, leaving vacancies behind them, leads to the gradual movement of atoms through the solid. This process is called diffusion. It proceeds faster as temperature is raised and can clearly lead to the mixing of two solids; for example, carbon will quickly diffuse interstitially in solid iron at high temperature, while a chromium layer, electrodeposited on steel, can be diffusion-bonded to the surface by a short heat treatment at 900° C.

The conduction of heat through a solid also depends on the vibration of atoms. If heat is supplied to the surface of the solid, the surface atoms become more agitated and transmit greater agitation to neighbouring atoms; this process continues gradually through the solid until atoms some distance from the heat source are vibrating more vigorously, representing a rise in temperature at this point. In metals and certain ceramics where free electrons exist in the atomic bonding mechanism, the electrons can acquire 'heat' energy by accelerating to higher velocities and so make a contribution to thermal conductivity. Free electrons, all carrying their negative charges, can equally transmit electrical energy readily; metals are thus good electrical and thermal conductors. Typical ceramics, with strong ionic or covalent bonds and few free electrons, have relatively poor electrical and thermal conductivity, the latter being dependent largely on transmission of atomic vibration.

The greater atomic vibration which occurs when a solid is heated has the effect of slightly increasing the mean spacing between atoms; this gives the well-known property of thermal expansion, the magnitude of which for many metals and ceramics is about 0·1% increase in length per 100° C rise in temperature. This is small but has important practical consequences, e.g. the buckling of a badly laid railway line on a very hot day.

The thermal agitation of atoms which leads to diffusion in metals and ceramics, and which sometimes assists dislocations to move through a metal crystal, shows that solids are far from 'dead' pieces of matter. It is this somewhat random agitation that permits new crystal phases to nucleate in a solid, if a change in temperature results in another crystal structure having lower energy and therefore greater stability. Such a change occurs in iron which at high temperature has the FCC crystal structure, but which on cooling changes to the BCC form at about 900° C.

The science which deals with the relation of thermal, mechanical and other forms of energy is thermodynamics. While the vibrational 'kinetic energy' of atoms plays a vital part in permitting change in a solid to a state of minimum total energy, we learn from thermodynamics that an 'ideal' change involving a large number of atoms never occurs fully; some energy is always lost or dissipated as heat. There is a degree of randomness favoured by nature, typified by atomic thermal agitation, and this results both in a balance between orderliness and chaos in solids, and in opportunites for change to occur to a more stable condition.

Much of what has been said about crystalline solids applies also to glass in which the atoms do not form an ordered array. The atomic structure of glass is more disordered and is like that of a liquid. It can be regarded as a molten ceramic which on cooling to room temperature, did not have time to form small crystal nuclei which would grow to form the usual polycrystal; instead, the liquid became more and more viscous until it resembled a solid more than a liquid. But not all ceramics can readily form glasses; of those that can, certain oxides are of most interest, particularly silica (SiO_2). This will readily form a glass by itself (fused quartz), but for many practical uses other oxides such as soda (Na_2O) and lime (CaO) are added to lower the melting temperature and improve various properties. In fused quartz, silicon and oxygen form strong bonds giving local three-dimensional order, but this does not extend to the stage of forming crystals. In common soda-lime glass the sodium ions (Na^+) and calcium ions (Ca^{2+}) take up somewhat random positions in interstices in a

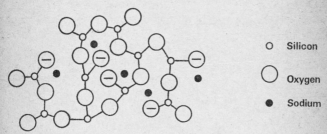

Fig. 8. *Structure of soda/silica glass. In pure silica, each oxygen atom is bonded to two silicon atoms and each silicon to four oxygens (only three are shown in diagram); the sodium modifies the silica network, fitting in near an oxygen ion bonded to only one silicon.*

modified silicon-oxygen network in such a way that local charge neutrality is maintained. Boric oxide (B_2O_3), like silica, forms a glassy network; in combination with alumina and barium oxide (BaO), it gives us glasses which resist corrosion by sodium metal and which are used for sodium discharge lamps. Glasses consisting largely of boric oxide and silica form the important borosilicate glasses which have a low expansion coefficient and are used for ovenware.

Most glasses are fairly good electrical insulators as they have no free electrons. Small electric currents can flow, however, due to the slow movement of ions, especially in the case of the rather mobile Na^+ ion. Ionic currents are very much smaller in a special lead silicate glass, containing no soda but a high lead-oxide (PbO) content. This glass is used to seal lamp filament leads where very good electrical insulation is required.

Glasses conduct heat essentially by transmitting atomic vibrations and, like most ceramics, are poor thermal conductors. They exhibit diffusion properties but, not being crystalline, have no dislocations or grain boundaries and have no facility for slip. Glasses are, of course, quite brittle at ordinary temperature and have low strength.

4. Microstructure – The Basis of Properties

The microstructure of a solid embraces several features: the nature of the phases present; crystal imperfections such as grain boundaries and dislocations; the purity of the solid in regard to unwanted chemical elements; any gross defects such as porosity or cracks. Much research in metallurgy and ceramics has the object of obtaining microstructures, whether of very pure single crystals or polycrystalline materials, which have special properties, e.g. strength at high temperatures.

Many microstructural features of a metal or ceramic can be seen in an ordinary optical microscope. The surface of the material is first ground flat, usually by rubbing on a series of emery papers and finally on a smooth cloth impregnated with a liquid suspension of fine alumina or diamond particles. The flat surface can then usually be etched by an acid or other chemical reagent so that grain boundaries and crystals of particular phases are attacked to various extents. When the etched sample is illuminated under the microscope the unattacked areas appear bright, while the grain boundaries and other affected areas do not reflect light so well and appear shaded to various degrees.

The microstructure of a stainless steel containing chromium, nickel and niobium is illustrated in Fig. 9. The main polycrystalline matrix appears white (unetched) and consists of iron with chromium and nickel in solid solution. The large particles at the grain boundaries consist of chromium carbides, (e.g. $Cr_{23}C_6$), while the fine 'black' particles within the grains are niobium carbide (NbC). The carbide crystals have been 'precipitated' from the iron-rich alloy by a process of crystal nucleation and growth during a heat treatment. Such precipitation usually gives rise to hardening and strengthening of an alloy.

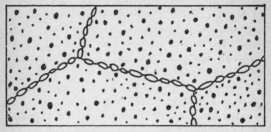

Fig. 9. *Microstructure of a Cr-Ni-Nb steel.*

The process of precipitation hardening is well illustrated by an aluminium-4% copper alloy. Fig. 10 is the phase diagram showing the regions of temperature and copper concentration

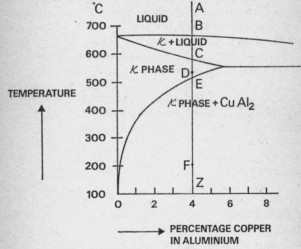

Fig. 10. *Aluminium-copper phase diagram.*

for thermodynamic stability (minimum energy) of the phases which may occur in the microstructure. If we follow the line

AZ, on cooling from the melt we find that, instead of solidifying at a single temperature, the alloy begins to freeze at B and is not wholly a solid polycrystal until temperature C is reached. Between C and E the alloy exists as a single phase solid solution of copper in FCC aluminium (the κ phase). But when the temperature falls below E (500° C) the aluminium-rich matrix is no longer able to retain all the copper in solution; thermodynamically, two phases now have a lower total energy and thus greater stability than one, so the copper precipitates, actually as particles of the compound $CuAl_2$. The crystal structure of $CuAl_2$ is different to that of the parent matrix, but one of its planes has similar atomic spacing to a crystal plane of the matrix. Particles of the new phase thus grow on the aluminium-rich FCC crystal planes and appear in the microscope as an oriented criss-cross pattern precipitate, the so-called Widmanstätten structure, first found in iron-nickel meteorites.

Besides considering the thermodynamics of microstructural change, it is desirable to look at the kinetics or speed at which change can occur. Before a new phase can be nucleated, the random diffusion mechanism has to provide a chance grouping of a sufficient number of the right atoms to form the new particle, with sufficient energy to spare to create the new crystal boundary. The process is thus time-dependent. For example, if the Al-4% Cu alloy is suddenly cooled from the point D by quenching in water, the κ phase will be fully retained in solid solution, because nucleation of the new phase has not had time to occur; furthermore, diffusion at room temperature is so slow that the frozen single-phase microstructure is temporarily stable. If subsequently the alloy is heated to point F (200° C) and held, or 'aged' at this temperature, a large number of $CuAl_2$ particles will slowly precipitate in accordance with the phase diagram. Such particles, at first not visible in the optical microscope, will strain the local matrix elastically and cause hardening. On further ageing, the particles grow; ultimately, the larger particles grow at the expense of the smaller ones, the atoms from the latter going back into solution and diffusing to the larger ones. This process reduces the energy due to all the local strain and crystal boundaries and results in softening again. The alloy is now 'over-aged', a form in which it is not usually required for practical purposes.

When an alloy solidifies from the melt there is normally a range of temperature over which different constituents freeze. This gives rise to an uneven distribution of alloying elements in the cast solid; the temperature gradients during solidification also produce an exaggerated crystal growth in certain directions, giving a 'cored' microstructure typical of a cast material. To convert this into a homogeneous polycrystal of 'equiaxed' grains, a metal can be hot-worked by forging or extrusion. This combination of deformation and thermally activated diffusion produces a more uniform alloy.

A plastically deformed (work-hardened) metal will itself recrystallise during a high temperature anneal to give new grains of lower strain energy and lower hardness. This occurs by nucleation, in regions of high elastic strain, of new crystals which eat up the old work-hardened microstructure. When a metal is recrystallised at a relatively low temperature, it usually produces a large number of nuclei and thus gives a small recrystallised grain size. If the temperature is too low, or the work-hardening insufficient to produce recrystallisation, a metal may still soften by 'recovery'; in this process local elastic strains in the crystals are relieved by thermally activated vacancy and dislocation movements. A recovery anneal often leads to dislocations forming new low-energy sub-boundaries, separating small angular differences in grains – a process known as *polygonisation*. In addition, a high temperature anneal, even in the absence of work-hardening, always produces continuous grain growth by the larger grains at the expense of the smaller ones; this is again energy activated since, for instance, a large grain has only half the total geometrical boundary area of eight small grains each one-eighth of its volume.

Grain boundaries, dislocations, foreign atoms and precipitates are all important because they greatly affect the strength of a metal or alloy and its ductility – the amount of plastic strain it can tolerate before fracture. It is possible with great care, to prepare needles or 'whiskers' of metals, a few microns in diameter, which are perfect single crystals containing no dislocations and, of course, no grain boundaries; their strength is exceptionally high, usually a few million psi, and they finally fracture, at several percent wholly elastic strain, by physical separation of planes of atoms against the strong bonding forces which exist. But normal metal crystals contain many dislocations which can move across the atomic planes at much lower stresses, e.g. thousands of psi. The continual slip and work-hardening that occurs as the dislocations interact with each other results in considerable ductility. A metal can often be stretched out to a very narrow cross-section before a crack is generated at some point of intense plastic strain.

At ordinary temperatures a metal is weakest, due to dislocation movement, when it is a pure single crystal or has a large grain size; a fine-grained metal is stronger, since the grain boundaries act as barriers to dislocations and strains are distributed more homogeneously. Work-hardening in a metal, imparted by plastic deformation, also strengthens it, since the dislocation tangles obstruct further dislocation movement. A heavily work-hardened metal, however, has less ductility left in it and so can crack more readily. Foreign atoms in solid solution, because of their different size, also impart a degree of strain throughout the polycrystal and so make dislocation movement more difficult. Many alloys rely on solid solution strengthening, but a much greater effect is obtained by fine precipitates, or even added ceramic particles, distributed throughout a polycrystal; these severely obstruct dislocation movement and greatly increase the yield stress at which slip and plasticity commence. Sometimes, however, the intense local strains created give rise to fracture cracks at low ductility when the applied stress reaches a certain value. In an alloy strengthened by work-hardening as well as by the solid solution or dispersed particle effect, softening by a high-temperature recovery anneal is also inhibited, since this too relies on easy dislocation movement.

Ceramic microstructures have many similarities to those of metals. Ceramics often form solid solutions with other ceramics and, according to their phase diagrams, form new phases of different crystal structure which nucleate and grow. But because ceramics are brittle they cannot normally be mechanically worked to homogenise them or to produce a fine recrystallised grain size. They are, in fact, not usually made from the melt, which would commonly be at a very high temperature indeed and would result in a very coarse-grained and weak material, probably not even of stoichiometric composition. Ceramics are frequently made by compacting fine powders and consolidating them by high-temperature sintering at well below their melting temperature. Diffusion both homogenises the microstructure and permits the porosity to escape. Each powder particle contains one or more grains, and the resultant polycrystal often has a quite fine grain size. Grain growth occurs in ceramics at relatively high temperatures, but we do not normally find mobile dislocations in ceramics.

For many pure ceramics it is difficult to sinter the powder to full density and thus avoid all porosity; in commercial aluminas a small glass content is intentionally added to assist sintering and ensure freedom from porosity. But several other and more novel ways exist for preparing non-porous and high-quality ceramics. Also, whiskers of ceramics can be prepared with strengths of several million psi, e.g. in alumina and graphite.

When stress is applied to a ceramic crystal or polycrystal, it undergoes elastic strain until a stress is reached where a crack is generated, perhaps at some macroscopic defect or grain boundary; the crack then propagates through the solid. Fine-grained ceramics are usually stronger than coarser-grained ones, as the local elastic strains in grain boundary regions are more evened out as grain size becomes smaller. In the case of glassy materials, it is believed that very small cracks (Griffith cracks) are always present at the surface and that fracture occurs when the applied stress is high enough to cause a crack to move or open out. However, it has recently been shown that a fine dispersion of insoluble particles introduced into a glass, can make crack propagation more difficult and thus strengthen

MICROSTRUCTURE – THE BASIS OF PROPERTIES

the solid. Typical fracture strengths of some ceramics and glasses, and strengths and ductilities of some metals are given in Table 3.

Table 3. *Typical mechanical properties of ordinary metals, ceramics and glass.*

Metal	Yield strength (*psi*)	Tensile (fracture) strength (*psi*)	Ductility (elongation at fracture)	Ceramic or Glass	Tensile (fracture) strength (*psi*)
High tensile carbon-steel (heat treated but with no alloy metal addition)	70 000	100 000	15%	Alumina (commercial quality)	35 000
80% Nickel-20% Chromium alloy (cold-drawn and annealed)	60 000	110 000	35%	Porcelain (commercial insulator quality)	15 000
Aluminium sheet (commercial purity – cold-rolled)	17 000	20 000	10%	Glass (soda-lime-silica or borosilicate)	10 000

The exact chemical composition of a metal or ceramic may have an important effect on its microstructure; sometimes sophisticated methods are used for the quantitative analysis of minor constituents. Much information, however, can be obtained by physical examination techniques. Optical microscopy can be used to determine grain size, or to reveal the presence of extensive plastic deformation or the sub-grains of a recovery microstructure. It can also give information about the phases present and can distinguish features of a few microns size. But optical magnifications above 1000 times do not resolve further detail as we are limited by the wavelength of visible light – about $\frac{1}{2}$ micron. If we use an electron microscope to look through a very thin specimen or to detect its surface detail, much larger useful magnification is possible –

up to 1 000 000 times. Using electron microscopes, in which electrons can be accelerated by an electric field of 100 kilovolts (or 1000 kV in instruments now becoming available), very minute detail such as dislocations can be seen and photographed. In fact, the limit of resolution is just a few Ångstrom units. In the 'field-ion microscope', a newer instrument in which gaseous ions are accelerated to a fluorescent screen from a crystal under examination, an image of actual atoms and crystal planes can be seen. However, a magnification of 50 000 with an electron microscope is often adequate in materials research work to supplement optical microscopy. Preparing thin wafers for transmission photographs is quite difficult, especially for ceramics; one advantage of the new 1000 kV instruments will be their ability to penetrate thicker, more robust specimens.

The basic method of determining what crystals are present in a microstructure is that of X-ray diffraction. The wavelength of X-rays is comparable to atomic diameter or crystal lattice spacing. On passing through a crystal, X-rays are bent or diffracted by an angle which is determined by their wavelength and the lattice spacing. It is thus possible to irradiate a polycrystal with X-rays of given wavelength and, by measuring the angle at which the diffracted images are produced on a photographic plate, to calculate the lattice spacing of parallel sets of atomic planes. We can then identify the crystalline phases fairly readily, especially if we have an analysis of the elements present.

Sometimes it is necessary to know the amount of a very minor element in a material, which may be present to a few parts per million or even much less. For this, a mass spectrometer can normally be used; this is an instrument which has recently been highly developed for impurity analysis. The material, if a solid, is vaporised and ions are projected by an electric field into an area between the poles of an electromagnet. The magnetic field deflects the moving ions in a curved path onto a photographic plate. As the ions have a single electronic charge, or a simple multiple of this, but vary in mass according to chemical composition, similar ions are deflected to arrive on a single line on the photographic plate. The spectrum of lines on the plate indicates the elements present and the relative intensity of each line gives the relative amount present.

Another modern instrument used in analysing microstructure is the electron probe microanalyser. If a high-intensity narrow beam of electrons is accelerated in an electric field onto the surface of a solid, some of the electrons already present in the atoms of the solid become 'excited' to higher energy levels, i.e. they jump out of their own shells, around the atomic nucleus, to outer shells where they are unstable. The electrons immediately fall back to the lower energy shells, but in doing so give up their energy as X-rays which are emitted and can be analysed for wavelength. The various energy level differences of the atoms are known, so the wavelength of the X-rays emitted reveals which chemical elements have been excited by the electron beam. A beam as narrow as one micron can be used, so that minute detail, such as a precipitate in a grain boundary, can be analysed. The results obtained are semi-quantitative, but it is now possible to carry the analysis a stage further and subsequently vaporise the feature of interest, to a small depth, and analyse the vapour more precisely with a mass spectrometer. Vaporisation is done either by a higher-intensity electron beam or a narrow pencil of high-intensity light using a 'laser' source.

If we want to study the microstructure and properties of materials in some detail and ultimately to develop new alloys and ceramics with special properties, much research has to be carried out on pure materials or simple alloy systems, in order to understand the inherent properties over which we want control. Pure single crystals have yielded much information; these require special preparation techniques and, in some cases, have themselves found important practical uses, e.g. single crystal silicon (see Chapter 9). Most practical applications, however, concern polycrystalline materials, and it is of interest to know some of the techniques used in their preparation in the laboratory.

To prevent oxidation or contamination at high temperature, alloys can be prepared by melting in a suitable ceramic crucible or container under a molten ceramic flux, but a better and more versatile method for the laboratory is to enclose the crucible in a chamber under a vacuum or inert gas atmosphere. Ordinary electrical (resistance) heating or induction heating can be used, the latter utilising a high-frequency coil to induce a self-heating current in the alloy. The molten alloy may then be cast, still in vacuum or an inert atmosphere, in a mould which allows rapid solidification to a shape that can be handled. It might then be annealed in an inert atmosphere in an electric

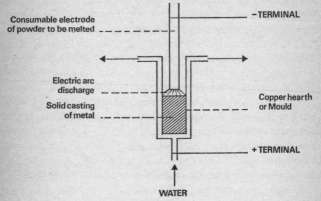

Fig. 11. *Arc melting of metal.*

furnace to homogenise it, then extruded or rolled to a form convenient for the examination of its properties. In the case of chemically reactive metals, great care has to be taken to avoid contamination by gases and other elements. A useful method of preparing the reactive refractory metal alloys is inert gas or vacuum 'arc melting' (Fig. 11). This utilises a water-cooled copper hearth, and an electric arc is struck between an electrode and the alloy on the hearth, melting the alloy. The copper hearth conducts the heat away adequately and retains a thin solid layer of the alloy in contact with it. The electrode may be of tungsten or may consist of the material to be melted, consolidated perhaps as an inhomogeneous mixture of powders of the metals concerned. If the latter is used (consumable electrode vacuum arc melting) the electrode is moved down as it melts and a casting is built up in the copper mould. Furnaces of this type are used for the commercial production of such metals as zirconium and titanium. In vacuum arc melting, many of the common gaseous impurities are drawn off and low melting-point impurities vaporised away. If a very pure metal or alloy is required, the highest quality starting materials must be used; many pure metals can be additionally purified by a zone-refining technique, as described for single crystal silicon in Chapter 9.

Polycrystalline ceramics are commonly prepared by first compacting a suitable quality powder or mixed powder of the constituents in a mould under pressure, then sintering the porous solid without applied pressure, at high temperature, below the melting point of the main constituents. The solid shrinks as the powder particles are drawn together to minimise their surface area; the pores, which usually contain gases, also become smaller as the gas diffuses from them and escapes, probably along grain boundaries.

Frequently some pores are left at the grain boundaries of the consolidated ceramic, while other pores may become trapped in new grains which have grown. The precise temperature and gaseous atmosphere of the sintering furnace is often important, as is the state of the initial powders, which must be prepared to the required purity and particle size, e.g. by calcining a pure chemical and milling it to fine particles.

The classical way of forming common ceramics, which contain glass on firing, is to shape a wet powder mixture of clay and other constituents. An extrusion technique is now often used for uniform shapes. In the case of fully crystalline and purer ceramics, two wet methods other than extrusion are of special interest, viz. slip casting and electrophoresis. For the former, a slip (liquid and powder mixture) is poured into a porous mould so that the liquid gradually drains away leaving the powder in a compact form for firing; in electrophoresis, the particles are finely dispersed as a suspension in a liquid and become electrically charged so that they migrate to one of the electrodes where they become compacted on a mandrel. The compact is finally fired to give an article of the desired shape, e.g. an alumina radar window (radome).

Another laboratory technique is pyrolytic deposition from the vapour phase. Carbon and certain compounds such as silicon carbide (SiC) and boron nitride (BN) may be produced as dense deposits on a heated surface by high temperature reactions by suitable compounds which vaporise readily. The method may utilise a single wire or tube as substrate for the deposit, or alternatively a large mass of particles may be coated in a chamber through which the vapour is blown – the so-called fluidised bed method.

As most ceramics cannot be mechanically worked, it is important to prepare them in the shape required and with the microstructure required. Some other special techniques will be discussed in Chapter 10, such as hot pressing of powders, which combines high temperature sintering with pressure, giving certain advantages. Whether metals or ceramics, however, purity and method of preparation are often vital to the microstructure required; and details of microstructure, such as grain size, precipitates or porosity, have gross effects on the material's properties.

5. Limitations of Metals and Ceramics

In engineering design it is important to make the best use of metals and ceramics. We may be forced to use a costly metal like tungsten (for a lamp filament), or some special ceramic for a rocket nozzle, but for normal purposes such low-cost materials as mild (low carbon) steel, aluminium, copper, porcelain and glass have many excellent properties. However, we would soon detect serious limitations in their strength, electrical properties, corrosion behaviour and in other properties if we attempted to use them outside a quite narrow range of conditions. The existing alloys and other materials which have been developed are constantly being found to be inadequate for new applications. It is then often necessary to develop quite exotic new materials, containing uncommon metals, and use sophisticated process techniques to obtain a desired range of properties.

Certain physical and chemical properties, however, are intrinsic to the various metals and ceramics and form a framework for research. These are listed below and will be considered in some detail.

Intrinsic properties of materials

Density	Thermal expansion
Elasticity	Thermal conductivity
Chemical stability	Self-diffusion
Melting point	Electrical resistance
Cohesive strength	Magnetic behaviour

The element of lowest density (weight per unit volume), ignoring those existing as gases at room temperature, is lithium (0·53 gramme/cm^3). As we go up in atomic number the atomic nuclei become heavier, but the atomic and ionic sizes do not increase much. The densest elements are the Group VIIIB metals, osmium, iridium and platinum (about 22 g/cm^3), while their neighbours in the periodic table, tungsten, rhenium and gold, as well as uranium, have densities over 19 g/cm^3. All the Group IA alkali metals are very light, but the lightest of the more common elements are the metals magnesium (1·7 g/cm^3) and aluminium (2·7 g/cm^3), and the ceramics, carbon and silicon (both 2·3 g/cm^3). Such light elements are of interest for applications in airborne equipment where low weight is necessary, but unfortunately they are not very strong mechanically. Iron and other metals used in high-strength steels (Cr, Mn, Co and Ni) all span the density range 7 to 9 g/cm^3. Titanium (Ti) is an inherently strong metal of lower density (4·5 g/cm^3), but of all the lighter elements, only carbon has significant strength at high temperature.

Elasticity governs the stiffness of a material or the amount it will bend when stressed. Young's modulus of elasticity has been mentioned as 30 million psi for steels. Its value is up to double this for some of the denser elements and refractory ceramics, but it is 10 million psi or less for the lighter materials, aluminium, magnesium and glass; the latter accordingly lack relative stiffness as structural members.

A material is normally required to retain chemical stability in its environment. In this respect carbon is rather limited since, although it retains its strength to 2500° C, it corrodes in air at 400° C to form an oxide gas. Even the oxidation-resistant metals, such as chromium, will oxidise significantly at 1000° C which is about the upper limit of use for metals without a special protective coating; ordinary steel and most metals oxidise to some extent at 500° C (dull red heat). Again, a metal may be stable at high temperature, yet form a brittle compound with another metal it contacts and so be incompatible with it; aluminium cannot be used to clad uranium for this reason. Zirconium, another nuclear cladding metal, slowly reacts with water or steam (H_2O) at 300° C, forming zirconium oxide (ZrO_2) on the metal's surface and liberating hydrogen; the latter dissolves in the zirconium, forming zirconium hydride (ZrH_2), and this embrittles the metal.

Electrochemical corrosion may occur when two different metals are in contact with a liquid such as rain- or sea-water; the less electronegative metal is attacked by anions in the liquid, while the cations present (i.e. metal or hydrogen ions) are attracted to the more electronegative metal (the cathode) where they do little damage. This type of corrosion can also occur when two phases in a microstructure have different electronegativity. It is often prevented or reduced by deliberately introducing an even less electronegative metal, which is preferentially corroded, e.g. a zinc coating on steel, by which the steel is made the cathode, and is protected from rusting, even if the coating is imperfect.

Ceramic compounds are often much less reactive than metals, some borides and nitrides being stable in air well above 1000° C and some oxides above 2000° C. Such compounds may decompose or vaporise as their melting point is approached. Other, less strongly bonded, particularly univalent, ceramics such as NaCl have low melting points and readily dissolve in water, their ions then losing their crystalline solid form.

The melting point of most metals is below 2000° C. Melting occurs when the cohesive or bonding forces, which maintain the ordered structure of a crystal, cannot resist disruptive thermal agitation of atoms tending to disorder the structure. Many strongly bonded ceramics, such as oxides and carbides, melt between 2000° C and 4000° C. Strong cohesive forces between planes of atoms in a crystal may also manifest themselves as high fracture strength. Fracturing a solid creates new external surfaces and thus extra thermodynamic energy (surface energy); this has to be supplied by the mechanical force necessary to rupture the atomic bonds.

Thermal expansion and thermal conductivity have important practical consequences. If heat is supplied to one part of a massive solid whose thermal conductivity is low, some time will elapse before the temperature of the cooler part rises very much. A heated surface, for example, tries to expand but is constrained by the cooler mass which expands less; this produces a 'thermal stress' which is compressive at the heated surface and tensile in the cooler region (Fig. 12). If heat is applied suddenly, the stress produced by such 'thermal shock' can fracture a brittle solid. This happens when boiling water is poured into a thick glass of soda-lime-silica composition. Borosilicate glass has much lower expansion and does not crack so readily, while fused quartz is even more difficult to crack by thermal shock. One ceramic, beryllia (BeO), is exceptional in that its thermal conductivity at moderate temperatures is better than most metals and 10 to 20 times that

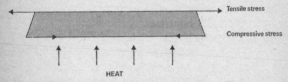

Fig. 12. *Thermal shock. Heat applied suddenly to a plate of poorly conducting material produces a tensile stress on the cooler face which may fracture it if it is brittle.*

of many ceramics. Some carbides and silicides have moderately good thermal conductivity and are thus less liable to thermal shock failure. Metals are good thermal conductors although they often have greater thermal expansions than ceramics; even severe thermal shock, however, does not lead to fracture unless the metal is brittle for some reason and therefore unable to deform plastically.

Self-diffusion, the process by which atoms move through solids by a series of thermal jumps, is much affected by the crystal defects present and by temperature; such diffusion gives rise to recovery and grain growth in metals. The rate at which diffusion speeds up as temperature is raised depends on the thermal 'activation energy' needed for an atom to jump to a new position. If the activation energy is large, the diffusion rate can increase by a factor of 100 for a small temperature rise, say from 500 to 550° C; even if the activation energy is small, diffusion may speed up by a factor of 3 or 4 over this

LIMITATIONS OF METALS AND CERAMICS

temperature range. Many high-temperature properties, such as oxidation rate and strength, are governed by self-diffusion, or diffusion of one element in another; this may lead to quite sudden property deterioration as temperature is increased.

Electrical resistance varies widely with materials. Typical oxide ceramics and glass have no free electrons and are insulators of very high resistance. Certain borides, carbides and silicides have a degree of free electron-bonding and are fairly good electrical conductors. Others, such as silicon (Si) or barium titanate ($BaTiO_3$) containing impurities of the 'wrong' valency, are 'semiconductors' of resistance intermediate between insulators and metallic conductors. These have important uses in electronics as described in Chapter 9.

All metals contain free electrons and are good electrical conductors; their precise resistances depend on details of structure, certain very pure metals having lowest resistance. Of the common elements, copper is the best conductor, aluminium second best while iron is relatively poor. Since the power which an electric current dissipates as heat is proportional to resistance, copper is a favoured electrical conductor; it is, however, relatively dense and expensive. Consequently, if the size of the conductor is not limited (although it often is in electrical equipment) we can use a thicker section of aluminium, which is a lighter and cheaper conductor giving equal power loss. Aluminium is, in fact, used for overhead power lines. Even sodium, which is a somewhat poorer conductor, but still lighter and cheaper, is now being used in underground cable; this chemically reactive conductor is encapsulated in a plastic sheath.

Copper and silver (the best conductor of all) are used widely as electrical contact materials for switches, but if very small currents must be switched, contacts made of gold or rhodium, which do not tarnish, may be necessary.

Impurities, alloy additions and work-hardening all increase the resistance of metals by straining the crystal lattice and impeding the flow of electrons. A common high-resistance alloy used for electric heaters is based on an 80% nickel-20%

chromium composition. Its resistance is ten times that of pure iron, nickel or chromium and it does not oxidise much at bright red heat (800° C).

As the temperature is lowered, there is less thermal agitation of atoms to impede electrons and the resistance of metals decreases. Approaching the 'absolute zero' of temperature ($-273°$ C, or $0°$ K) many pure metals become 10 000 times better conductors than at room temperature, while certain metals, alloys and compounds exhibit a remarkable property of 'superconductivity' whereby they lose all their resistance. Up to now, this low-temperature property has found limited but scientifically important applications, as mentioned in Chapter 11.

The principal magnetic metals are iron, cobalt and nickel. Some alloys and compounds are magnetic, and certain rare earth elements are magnetic at low temperatures. Magnetism has its origin in the spin of electrons in sub-shells of atoms of certain B-group elements. It arises where atoms in a crystal, to achieve minimum energy, have an unequal number of such electrons spinning in the two possible directions. The macroscopic solid can then be magnetised, by lining up the 'domains' of the solid which have their net spin in a particular direction; this gives the solid a north and south magnetic pole. A north magnetic pole then attracts another magnetic solid's south pole (which may be induced by the first magnet) since energy is minimised by their close proximity or contact. Iron is the most important magnetic material and can be magnetised to greater intensity than almost any other material; nickel-iron alloys can be magnetised extremely easily, while certain ceramics of spinel crystal structure are important magnetic materials by virtue of being also electrical insulators; they can, for example, be wound by an uninsulated electrical conductor or used at a high frequency with very low power loss. Magnetic materials find many uses in electrical equipment and in electronics.

Besides the intrinsic properties of materials, other properties which relate to mechanical behaviour are important. These are listed below.

LIMITATIONS OF METALS AND CERAMICS

Properties related to mechanical behaviour

Yield stress	Creep properties
Ductility	Microstructural stability
Brittleness	Fatigue properties

When a stress is applied to a bar of polycrystalline metal, e.g. an age-hardened aluminium-copper alloy, a strain occurs as illustrated in Fig. 13. As the stress increases to 20 000 psi

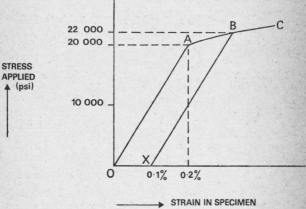

Fig. 13. *Stress–strain curve for Al alloy*.

(the yield stress), the strain is wholly elastic (0·2%) and follows the straight line OA. Beyond this stress, dislocations in the crystal are activated and slip occurs on the various planes giving plastic strain indicated by the line AB; the slip-line markings on the metal's surface can be seen through a microscope. If the stress is now removed, the bar will contract elastically along the line BX, parallel to OA, so that a plastic strain or work-hardening represented by OX will remain in the metal. If OX is 0·1% strain, the stress at B is called the 0·1% yield stress (22 000 psi in our example), which is slightly higher than the initial yield stress at A. If the stress is reapplied, the strain in the specimen follows the line XB and continues to increase by plastic deformation along BC with slowly increasing stress. Eventually a maximum stress is reached and the metal fractures, usually with many percent plastic strain. The ductility of the metal is measured either by this percentage elongation of else by the percentage reduction in cross-sectional area of the metal bar where it fractures; the latter may be over 80% since a ductile metal forms a neck before fracture. Metals and alloys differ very widely in their yield and fracture strength and ductility.

Brittleness or its opposite, toughness, is measured by the energy required to fracture a standard-size bar of metal usually

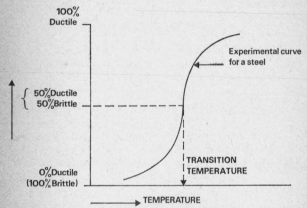

Fig. 14. *Brittle/ductile transition temperature.*

containing an artificially introduced V-notch. A heavy pendulum is allowed to strike the metal so that it rapidly deforms at the notch and perhaps generates a crack which travels along 'weak' crystal planes and rapidly fractures the bar in a brittle manner. The proportions of the fractured surface which are respectively brittle and ductile (of deformed fibrous appearance) can be readily observed. In BCC metals (notably carbon steel), which are limited in regard to easy slip planes, the proportion of the fracture surface which is ductile decreases as temperature is lowered (Fig. 14); the toughness or energy to produce fracture similarly falls. The temperature at which the fracture is 50% brittle and 50% ductile is called the transition temperature and represents a safe minimum working temperature if brittle fracture is to be avoided. This point is particularly important for welded carbon steels, which may contain porosity or even small cracks which act as notches. The classic failures by brittle fracture occurred in welded American Liberty ships during the 1939–45 war. A transition temperature as low as possible is now given to structural steels, principally by means of a very fine grain size.

Ceramics and glass are normally quite brittle and possess negligible ductility; their stress–strain 'curve' is a straight line up to the point of fracture. It is this feature of ceramics and glass which limits their use as engineering materials; a metal, on the other hand, if subjected to thermal shock or inadvertently over-stressed, can yield a little without fracture.

So far we have considered the stress–strain curve at 'room temperature'. At high temperatures certain polycrystalline ceramics, e.g. magnesia (MgO), will exhibit crystalline slip and thus yield, giving a moderate percentage of ductility before fracture. Metals also soften with rise in temperature and their yield strength usually falls rapidly at about half their melting temperature. This effect is associated with the increasing thermal agitation of atoms which assists the occurrence of slip; the grain boundaries of a polycrystalline solid are also affected in that, as temperature rises, they cease to rigidly 'cement' the grains together; they become less viscous and permit the grains to slowly slide past each other as the grains adjust their shape by processes such as slip and diffusion.

An important additional factor at high temperatures (i.e. at temperatures approaching half the melting point and above) is known as creep. This is a steadily increasing plastic strain which occurs with time when a constant stress is applied to a solid. Creep deformation may be produced by a stress above or below the yield stress; the rate of creep strain (percent per hour) increases as either stress or temperature is further raised; ultimately, a material may fracture when a certain 'creep ductility' is reached. Creep is clearly an insidious limitation in engineering design of high-temperature components which are stressed in service.

Fig. 15 represents a typical creep curve, the various portions of which can be regarded as due to different mechanisms of deformation. OA represents the instantaneous yield which occurs if the stress applied is above the material's yield stress; the primary or transient creep stage (AB) is usually due to slip in the grains activated by the combined effect of stress and thermal agitation. This then gives way to a constant rate secondary creep (BC) which may last for some time and may be regarded as due to a balance between work-hardening and thermally activated recovery, occurring together. These processes can occur in single crystals, but in polycrystals an

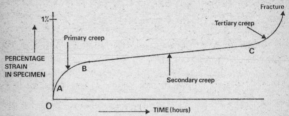

Fig. 15. *Typical creep curve*.

additional factor in secondary creep is the viscous grain-boundary sliding effect. The latter may ultimately give rise to small grain-boundary cavities or cracks at points of high stress; if so, the material will weaken and will commence tertiary creep which leads to separation of the grain boundaries in fracture. Such a brittle fracture, giving low 'creep ductility' of perhaps a few percent, is typical of material which has undergone creep strain very slowly for thousands of hours. Clearly, the knowledge of the stress and temperature which a material can withstand without excessive deformation or failure over many years is most important for engineering design.

Many alloys have been developed specifically to possess high creep strength. They often contain dispersed particles of a second phase which block grain boundary and dislocation movement and raise the recovery temperature. Unfortunately, most strong materials lose their microstructural stability at high temperature; they may slowly degrade in strength because

LIMITATIONS OF METALS AND CERAMICS

a finely dispersed precipitate over-ages giving coarser particles, or because grain growth occurs enabling newly positioned grain boundaries to avoid precipitates which 'lock' them. At temperatures where these processes slowly occur, the rate of creep increases, in the manner of tertiary creep, and the material in due course is likely to fracture.

Ceramics exhibit creep generally at much higher temperatures than metals. At 1000–1500° C magnesia behaves much as steels do between 400° and 700° C, or magnesium or aluminium between 100° and 400° C. Many ceramics, such as polycrystalline alumina, cannot creep very much by crystalline slip even at very high temperature, but they can deform under stress by permitting diffusion of vacancies to occur. This results in a bulk deformation effect and occurs also in metals at low stresses at temperatures approaching their melting point; it sometimes results in very large ductility or 'superplasticity'. Glass, with its characteristic structure, does not creep like a crystalline solid, but behaves as a viscous liquid giving a constant rate of plastic strain which increases strongly with temperature, for a given stress. Such deformation is essentially a stress directed self-diffusion process. Soda-lime-silica and borosilicate glass begin to flow very slowly at about 600° C, while fused quartz (silica) is relatively rigid up to much higher temperatures.

Fatigue, another limiting property, may occur when a stress lower than the fracture stress is continually removed and reapplied, or even reversed. Ultimately such stress-cycling can produce fracture, which is due to accumulation of repeated small deformations producing local damage to the material's microstructure, resulting in a crack. If the fatigue stress is relatively high, say just above the yield stress, so that measurable deformation occurs in the first few cycles, failure is likely after relatively few cycles, perhaps hundreds or thousands. This type of fatigue is called 'high-strain fatigue' and sometimes results from repeated thermal shocks, e.g. in certain turbine components which are heated up daily to meet the varying electricity load. Failures can also occur by a combination of creep and fatigue.

If the fatigue stress is below the yield stress of a metal, say between one-third and half the fracture strength, failure may still occur but after many millions of cycles. This is 'low-strain fatigue', the metal itself showing no measurable strain prior to fracture. Vibrational stresses such as occur in rotating or reciprocating machinery sometimes give this type of failure. A safe alternating stress for design purposes, provided creep is absent, is about 40% of the fracture strength, for steels. It is lower for some other metals and is always reduced if a mechanical notch is present which locally magnifies the stress. Fatigue can also occur in ceramics.

The selection or development of a material, for use in a given environment at a certain stress, temperature and with other peculiar conditions which may apply, clearly can present many problems!

6. Strategy and Tactics of Materials Research

In the last decade or so, research on materials has been vital to the development of nuclear reactors, high-temperature turbines, miniature electronic devices and modern aircraft and space vehicles. The materials scientist, concerned with developing new alloys and ceramics, has been involved both with basic science and with new concepts of engineering design which utilise new materials. In practice, science and its applications feed each other and, in combination, are the essence of technology. A new engineering concept, itself probably stimulated by a social need or by 'competition', often demands not only design and manufacturing skill, but a material with some properties beyond those at present obtainable. In other words, the existing materials and their properties have provided the framework for invention and engineering design in the past, but a more adventurous design or totally new device may be possible if a new alloy or ceramic were 'discovered', e.g. one which was stronger or did not corrode at high temperature or had lower electrical resistivity. This demand for a 'better' material provides stimulus to research with the particular objective in mind.

Equally, those directly engaged in research on metals and ceramics, whether attempting to understand basic atomic or crystal properties or to control them to meet a need, inevitably produce fresh information and discover new aspects of their materials; this may both add to scientific understanding and perhaps provide the basis of a technological breakthrough.

Research, invention and innovation often take unexpected turns and, by virtue of the necessarily unfettered intellectual content involved, cannot be controlled in the way that industrial production is controlled. Yet there has to be a policy framework for research in general, and for metals and ceramics research in particular, if only because millions of pounds are being spent and useful results are expected.

In Britain, research in materials science and engineering is carried out in three types of institution: universities, government laboratories and industry – although strictly, the laboratories of nationalised industries and research associations do not fit into just one of these categories. University research derives its funds from the Government, from charitable institutions and from various sponsoring bodies; here, basic science is pursued in depth on a variety of subjects, some specifically requested by sponsors, but most university research is only loosely co-ordinated from the point of view of any national science policy or strategy. The academic freedom of such research often results in excellent basic work in metal and ceramic science, which broadens and deepens our knowledge of the subject.

Research in government and industrial laboratories, although much may be quite basic, is related to particular objectives or special fields of interest. Such research, and the product development which follows it, may consist of 'large-scale' or 'small-scale' projects – the former often require the expenditure of millions of pounds, the latter only perhaps ten to some hundred thousands of pounds. The large-scale project, which may be concerned with a new type of nuclear reactor or aircraft, acquires a national status and a characteristic impetus and procedure; it often involves considerable government financing and several teams of scientists in different specialist laboratories working towards its objective; a significant proportion of the effort may be in metallurgy and ceramics, and of this, some will be probing the frontiers of knowledge while some will be applied science and development. Often, research initiated by industry itself has led to a major step forward, for example, the development of high-temperature steels and alloys which have revolutionised turbine power plant; nowadays, the industrial laboratory participates rather more in government sponsored large-scale research, particularly in fields where the firm may later assume a manufacturing responsibility.

Small-scale research, instead of being geared to a major national target, is aimed at developing new or better marketable products or more economically made products. Much industrial research is of this type, and here the strategy is concerned with fitting the 'correct' small-scale research projects into a company's corporate policy of future fields of interest and profitability. Clearly, careful selection must be made of lines of research, which should have a reasonable chance of leading to new products, better design or cheaper manufacture, and which accordingly will have a large impact on the market.

The cheaper generation of electricity by means of nuclear reactors is an example of very large-scale research, where Government expenditure on research and development has been at the 100 million pound level, some tens of millions having been in metals and ceramics research; considerable sums have also been spent on research and development in this field by private industry. Such large investments are justified if the rewards are commensurate; in this case the total cost of British nuclear power stations, built or contemplated, is approaching 1000 million pounds, and the goal of much cheaper electricity is now being realised.

After the initial discovery of nuclear energy release, research in Britain on this subject gathered pace in the late 1940s and the early 1950s. The initial tactics were to find out more about every aspect of the subject, and to demonstrate the feasability of controlling the release of heat in an atomic pile or reactor. Many aspects were being studied in parallel, one of the most important being materials research. Much coordination of the different researches, in relation to engineering design concepts, was necessary to ensure efficient progress.

When the first relatively small atomic piles had operated successfully, though not for electrical power generation, the engineering feasibility of a larger nuclear reactor system was considered in detail, as a realistic step towards commercial power generation. There were research problems on a number of subjects where feasibility had to be proved for different aspects of the proposed design. In the field of metallurgy, detailed properties of prospective alloys were examined, and materials with adequate properties under appropriate test conditions were developed within a necessary time-scale. Under such conditions there clearly had to be give and take between the expected performance of materials and components and the engineering design in which they could be utilised. In due course many feasibility exercises had shown in adequate detail that the building of a large nuclear-power reactor could be successfully undertaken. This was the Calder Hall plant, which, besides demonstrating large-scale electricity generation from uranium fuel, was also used to produce, in some quantity, the valuable artificial element plutonium (Pu) of atomic number 94.

When industry was asked for the first commercial tenders for nuclear power stations, the feasibility of their successful operation was beyond doubt, but the cost of electricity generation turned out to be about 1d per unit. This was dearer than the then best conventional coal-fired power stations, which themsleves had been improved by utilising steam at higher temperature and pressure (565° C and 1500 psi), and by building larger turbine generators of 100 to 200 megawatts power; these produced electricity for between 0·6d and 0·7d per unit. Ultimate generation costs had been considered throughout the nuclear-power exercise, and the whole project pursued in the reasoned view that costs would become at least competitive with electricity produced from other fuels; it was also believed that electricity consumption would expand so considerably that nuclear power would be a necessity, and that it was therefore important to build up a nuclear industry which would progressively produce more efficient plant.

Much research has been done and increasingly efficient nuclear power stations have been designed during the last 10 to 15 years; research on metals and ceramics has been prominent and has been carried out partly by the U.K. Atomic Energy Authority and partly in industrial laboratories. This work was well co-ordinated with the various objectives which arose; some details of the research and its results are described in Chapter 8.

The effort to design more efficient nuclear power stations has brought prospective generation costs down to between 0·4d and 0·5d per unit; this has stimulated research and development on competitive energy sources (coal, oil and natural gas) and means of generation other than by the large power stations. Modern large steam-turbine generators have cut prospective costs for conventional generation from coal to a little over 0·5d per unit, but these large turbine generators are used by nuclear plant, and so reduce the cost of nuclear power also. In the so-called 'fast' reactor systems of the 1970s, costs may approach 0·3d per unit and it appears likely that nuclear power stations will win the race as electricity producers. There remains, however, the formidable competition of cheap natural gas as a source of heat, both locally and possibly at central power stations for electricity generation. The strategy of future research on 'energy' must clearly derive from a national fuel and power policy; having decided this, the content and tactics of further research must be carefully tailored to meet feasible objectives in the most efficient manner.

The tactics of large-scale research may be summarised as: a perceptive appreciation of the results of basic research and adequate further studies on relevant aspects of this; a periodic analysis of technical feasibility and economic worthwhileness of objectives; and co-ordinated versatile applied research to prove or disprove important aspects of the project. While a project appears worth pursuing, it is obviously important to cut out unfruitful avenues of expensive applied research as soon as is possible, so that development can proceed with minimum wasted effort to a worthwhile conclusion. Ideally, the latter should also be a springboard for further technological advance.

Many of these tactics are also applied to small-scale research and can be well illustrated by a recent piece of research in industry which resulted in a new and superior type of carbide tool-tip for machining metal. Most of the cutting tools used, which have a high wear resistance, have consisted of a sintered mixture of tungsten carbide (WC) powder and a small amount of cobalt (Co) metal 'binder', often with additions of other hard carbides. Such tips have a good but limited life when machining steel, and are fairly tough in that they do not suffer from surface chipping. It was clear, however, that a more wear-resistant tip, which was tough and not too expensive to manufacture, would be a most desirable addition to the carbide-tool market. The size of this market is such that research and development expenditure in the tens of thousands of pounds would be appropriate if a superior product could be made.

Tungsten carbide/cobalt tips containing up to 20% titanium carbide (TiC) were already widely used for steel machining. TiC is harder than WC and imparts added wear resistance; it also has the effect of reducing the loss of WC due to the latter dissolving in the hot steel swarf removed by the cutting action. Unfortunately, greater TiC contents in WC/Co materials lead to increasing brittleness and aggravate the problem of surface chipping of the tip. One solution to obtaining a better steel-cutting tip of adequate toughness has been to add other, more expensive, carbides such as tantalum carbide (TaC). This results in a rather costly product and it would be better if tips utilising a larger amount of the inexpensive TiC could be made, with less brittleness. High TiC content materials had in fact been produced, but were too brittle to be widely used, despite their excellent wear resistance.

In brief terms, this was the technological background of the project. There was in existence a considerable basic knowledge on the sintering of carbides, their compatibility with binder metals, and details of microstructure and properties. Tungsten carbide has an appreciable solubility in titanium carbide and it was thought that a more thorough examination of mixtures of these two carbides might be worth while. It was surprisingly found, while examining a wide range of TiC/WC compositions with certain binder metal contents, that much less brittleness occurred than expected when about equal proportions by weight of TiC and WC were present. Such a material also had very good wear resistance. Concentrated work with approximately equal content TiC–WC mixtures was then carried out, and other aspects of the project were curtailed. The most suitable binder metal was found to be a nickel-molybdenum-cobalt alloy, and best results were obtained when a relatively large amount of binder was used for the tip.

The precise composition and sintering conditions for the new material were optimised, and detailed tests made of its properties and reproducibility. Microstructure, hardness, strength and practical machining behaviour were all examined at various stages of the project. Finally, field trials in a number of factories on different types of machining were carried out before the product was marketed generally. The new material, in fact, combined much of the best aspects of both TiC- and WC-based materials; the tip had a high TiC content by volume, due to the relatively low density of TiC, and possessed the good wear properties characteristic of this carbide; it also possessed toughness or resistance to chipping more characteristic of the conventional WC-based materials. The microstructure giving these properties consisted of a large content of strong ductile binder alloy surrounding rounded particles of TiC/WC solid solution, a few microns in diameter. There was insufficient WC present to saturate the TiC and form another phase, which would consist of angular WC particles, but there was sufficient WC to reduce the brittleness of a high TiC-content material.

In the above example of industrial research, the tactics were first, an appreciation of the basic knowledge and experience in the subject and the linking of these with a desirable and perhaps feasible objective; secondly, defining and implementing a sufficiently broad research programme to encompass and test research ideas; thirdly, concentration on the most promising line of research to the point where one approach indicates feasibility of a worthwhile objective, which itself could be modified to make the best use of research results; fourthly, applied research to optimise all aspects of the material and its preparation, and a full programme of tests to prove a prototype manufactured product.

During the early exploratory stages of a small-scale project, the cost of research is not large, but as an objective becomes indentified and more specific lines of intensive research activity are envisaged, it is important to estimate the total likely cost of the project and the benefit to be derived from its successful completion. Such factors must be weighed with sufficient far-sightedness, and sometimes an element of intelligent gambling, at various stages of the work to decide whether continuation of a particular line of research is worth while.

7. Metals in Gas Turbines and Steam Turbines

Many research problems have been concerned with the satisfactory behaviour of metallic alloys at as high a temperature as possible; the extent of success achieved, particularly for highly alloyed steel and nickel-based alloys between 500° and 1000° C, has been a major factor in the development of efficient and powerful jet engines and steam turbines. Before these developments there were few applications where metals were used above 500° C – some examples are electric heater windings, furnace parts, tool steels for shaping hot metals, precious metal crucibles, and tungsten lamp filaments.

Below 500° C many steels can be used satisfactorily, but the mechanical properties of copper, aluminium and magnesium alloys deteriorate significantly if the temperature is raised to even 200° C; for important long-life applications, such as the Concorde aircraft skin structure, an upper safe limit for a strong aluminium alloy is little over 100° C.

The gas turbine, in simplest terms, consists of a rotor shaft to which a number of rows of radial blades is attached; oil is burnt in a combustion chamber and the hot high-pressure gas is directed over the profiled blades to spin the rotor on its axis. For jet aircraft use, the main function of the turbine is to drive an air compressor, mounted on the same axis, which supplies high-pressure air to the combustion chamber; the remainder of the energy in the high-velocity combustion gas is used to propel the aircraft forward. In the industrial gas turbine, many rows of blades are employed, and the whole of the energy available in the gas is used to spin the rotor; this drives the compressor and also a generator to produce electrical power. For marine use, the generator may be replaced by a geared driving shaft. In all cases the blades and rotor structure experience considerable stresses due to rotation.

The properties of metals required for the aircraft and industrial gas turbines are not quite the same. In the jet engine the life required of the turbine blades, and some other components, need be only some hundreds of hours; it is also economic to use a high-quality distillate fuel, the ash from which does not much corrode or erode the blades. But the industrial gas turbine for electricity generation should last many tens of thousand hours before overhaul, and should ideally utilise a cheap oil, the larger amount of ash from which may corrode the blades. These requirements raise two of the principal problems in high-temperature metallurgy – corrosion resistance, and adequate mechanical properties over a given period of time (especially creep).

Although similar alloys are available for each, the jet-engine blade temperature may be over 900° C for short periods, while the steady temperature of industrial gas turbine blades may be limited to about 650° C. One reason for the lower permitted temperature is the corrosive effect on the blades of vanadium pentoxide (V_2O_5) in the fuel ash, which melts at little above 600° C. This effect can be minimised in various ways, one of which is to add a silicon compound to the fuel; this results in a higher melting-point vanadium compound. If, however, a higher grade distillate fuel is used, the industrial gas turbine is able to run at a slightly higher maximum temperature.

Alloys based on the three Group VIIIB metals, iron, cobalt and nickel, often with a 15–20% chromium addition, have been developed to produce high-strength, creep-resistant and corrosion-resistant alloys for the blades and rotor body; the latter frequently consists of a series of forged steel discs, welded together. Very simple alloys based on iron, cobalt or nickel have poor creep properties above 500° C although they may tolerate higher temperatures from the point of view of corrosion. But alloys containing several metals in solid solution, with minor additions of other elements to give precipitates which add further strength, are capable of much higher working temperatures.

Steels containing significant amounts of nickel, chromium, cobalt, often with minor additions of molybdenum, niobium and other elements, rely partly on precipitates of intermetallic compounds and carbides to give high-temperature strength. Such 'steels' may contain only about 50% iron and their crystal structure is FCC, a form in which pure iron exists above 900° C. Other, less alloyed, steels are based on the normal BCC iron crystal structure and have the virtue that they can be 'heat-treated' by cooling at a suitable rate from the FCC structure through the phase transformation; they can then be tempered at a lower temperature to form a new microstructure. The 'austenitic' steels, which do not have this phase transformation and are FCC at room temperature, are sometimes work-hardened a little to strengthen them, but they can be recrystallised only after excessive work-hardening; this is quite difficult to accomplish in a large mass of strong steel. Such steels obtain their precipitation strengthening by an age-hardening treatment. Suitably alloyed steels have good long-term creep strength, yield strength, and ductility properties up to 600–700° C and have been used as gas turbine discs. They may contain 15% or more chromium to confer resistance to oxidation and corrosion.

The development of nickel alloys of increasingly greater creep strength has been very successful. The British 'Nimonic' alloys contain about 15–20% chromium with a small amount of titanium and aluminium; the versions with greater creep strength also contain significant amounts of cobalt and molybdenum. Again, a combination of solid solution strengthening with precipitation of compounds within grains and at grain boundaries provides the necessary properties. These alloys have a FCC crystal structure; their thermal treatment usually consists of a high temperature 'solution' anneal, followed by a somewhat lower temperature treatment when precipitation effects occur.

Nickel alloys have been widely used for aircraft gas turbine blades where they can withstand stresses at 900–1000° C;

frequently such blades are cooled by ducting cooler gas through channels in them. It has recently been claimed in the U.S.A. that single crystal gas turbine blades have been made in a nickel alloy, so that creep due to grain boundaries at high temperature is avoided.

Chromium, by virtue of a natural protective layer of chromic oxide (Cr_2O_3) which adheres to its surface, appears an obvious choice as a basis for a gas turbine blade alloy. Unfortunately, this BCC metal is embrittled by a trace of nitrogen which raises its brittle/ductile transition to well above room temperature. Nitrogen is difficult to remove when preparing the metal and can re-enter it if the metal's temperature is raised to a few hundred °C in air. One way of minimising embrittlement is a small addition of yttrium which results in a protective coating of the double oxide, yttrium chromite ($YCrO_3$). Other alloying additions to chromium have been tungsten, tantalum, titanium, silicon and boron; the best chromium alloys are claimed to be as good or better than nickel and cobalt based alloys and suitable for use as gas turbine blades at 1000–1100° C. Blades with still higher temperature creep strength are obtained with niobium-based alloys; these are discussed in Chapter 10.

In the modern steam turbine, the maximum temperature is usually limited to 565° C. High-pressure, high-temperature steam is supplied by a boiler in which pulverised coal or oil is burnt to heat tubes through which the high-pressure water flows; the steam produced is then superheated in other tubes, and delivered at 565° C through large pipes to the turbine valves. Equally, such steam may be generated in a heat exchanger by extracting heat from the primary coolant of a nuclear reactor. The steam enters the high-pressure (H.P.) cylinder or casing which contains the bladed rotor shaft; it expands through successive rows of stationary nozzles (in diaphragms attached to the cylinder) and alternate moving blades, fixed to the rotor (Fig. 16). From this labyrinth, the steam enters one or more low-pressure (L.P.) or intermediate-pressure (I.P.) cylinders where further expansion occurs past further sets of nozzles and blades. Towards the end of the L.P. cylinders, the cooler steam partly condenses and contains many water droplets before it finally passes at very low pressure to the condenser plant. The efficiency of extracting energy from the steam is now commonly increased by interposing one or more intermediate pressure turbines between the H.P. and L.P. turbines, and reheating the steam in the superheater between the H.P. and I.P. stages. The I.P. rotor and blades are necessarily larger than those of the H.P. stage, since they pass a larger volume of steam; as this steam had been reheated, probably

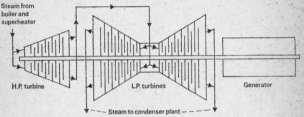

Fig. 16. *Simple steam turbine diagram.*

to 565° C, the metallurgical problems of the I.P. stage are those of temperature and large size, with corresponding large stresses.

Frequently, the H.P., I.P. and L.P. turbines and the electrical generator are mounted on the same axis; the generator is driven at 3000 r.p.m. with a truly vast power. The largest size units today deliver 500 and 660 megawatts of electricity and employ maximum steam conditions of 565° C and 2300 psi pressure: such equipment has provided a set of metallurgical problems beyond those solved for the first 'high-temperature' turbine generators, built some 15 years ago and employing steam at 565° C and 1500 psi, to deliver 60 MW.

The thermodynamic efficiency of steam turbines has not yet been increased by raising the temperature beyond the 565° or 600° C level, because this is the limit to which low cost, non-stainless steels have been developed with adequate properties in the form and size required for the bulk of components. The H.P. and I.P. cylinders, rotors and blades are stressed by the steam pressure or rotational forces or both; the close dimensional tolerances of the engineering design are such that creep strain in such components must be limited to a few tenths of 1% during their service life, perhaps lasting 20 years (175 000 working hours).

It is interesting to trace some aspects of low-alloy steel development which has provided us with present-day rotor forgings, superheater tubes, steam pipes and large cylinder castings with the high-strength bolts they require.

Ordinary carbon steels, of 0·2–0·5% carbon content, can be heat treated to give a range of yield strengths, increasing with carbon content; such steels have an excellent combination of properties up to 300–400° C, but at higher temperatures their creep strength deteriorates rapidly. It is also difficult to weld the higher carbon steels unless special precautions are taken to prevent the weld having a hard and brittle microstructure on cooling. If a steel is heated to about 900° C, the carbon dissolves completely in the FCC iron or austenite. On subsequent slow cooling, the structure recrystallises to the BCC structure in which very little carbon can dissolve; it precipitates again as iron carbide (Fe_3C), which can be made to appear in various forms and which imparts a degree of strength. If the cooling is fast, for example by quenching, an intermediate crystal structure forms which is very hard and strong but brittle; this can be aged or tempered, say at 500° C for a short time, to enable a fine distribution of carbide precipitates to grow. Such hardened and tempered steels usually have a fine recrystallised grain size and a good combination of strength, ductility and toughness.

When a large thickness of steel is cooled by water-quenching, the outside cools much faster than the centre of the mass; if it is a plain carbon steel, the inside will be softer and of moderate strength while the steel near the surface will be fully hardened and could be tempered to become strong and ductile. In practice, small amounts of other metals, such as nickel and chromium, are added to carbon steel which permit the hardening to occur at a slower cooling rate; oil-quenching or even air-cooling can then satisfactorily harden a large turbine rotor, which can subsequently be tempered to give the best combination of strength and ductility.

A nickel addition to steel dissolves in the iron polycrystalline matrix and increases strength and toughness. The nickel also slows down diffusion at high temperature and so reduces grain growth and consequent embrittlement if the steel should be overheated during heat treatment. A nickel steel, however, may deteriorate during long service at high temperature due to decomposition of the iron carbide (graphitisation). This can be counteracted by a small chromium addition which itself forms stable carbides and also remains partly dissolved in the matrix; both these factors contribute to better mechanical properties. A 3% nickel-1% chromium steel, containing about 0·3% carbon, will readily harden by air-cooling or oil-quenching (depending on thickness) and can be tempered to give a yield stress of over 100 000 psi with ductility of 20% elongation at fracture.

Nickel-chromium steels unfortunately possess poor creep strength above 450° C, and it is necessary to use up to 1% molybdenum in steel, or, better still, molybdenum and vanadium, if good creep strength is required at 450–550° C. The molybdenum partly dissolves in the iron and partly forms carbide precipitates, such as Mo_2C; it slows down diffusion and so resists over-ageing and loss of strength at normal tempering temperatures. A mere 0·3% of vanadium, a strong carbide former, dramatically improves high temperature strength; when a steel containing molybdenum and vanadium is suitably cooled from 900–1000° C and tempered for a short time at 650° C, the carbides notably V_4C_3, form very fine precipitates which impart high creep strength and stabilise the microstructure for long periods at 550° C. A low-carbon 1% Mo-0·3% V steel can withstand a working stress of 6000 psi at 550° C for 10–20 years with only 0·5% total creep strain.

Although Mo-V steels, especially up to about 0·8% V content and with little other alloying additions, have excellent creep strength, they tend to fracture during creep with very little ductility, sometimes as low as 1 or 2%. This is a most dangerous engineering property, since a very low secondary creep strain rate may suddenly turn into tertiary creep and fracture. High creep strength and creep ductility tend to be favoured by opposing factors which need to be carefully balanced to obtain the best overall properties. A large grain size with fine precipitates throughout the grains and at grain boundaries normally give greatest creep strength; the number of grain boundaries sliding is much smaller than in fine-grained material, while dislocation movements, slip and grain boundary sliding are all impeded. But if the grain is too hard, slight sliding at the weaker grain boundaries produces local high stresses which cannot be relieved and a crack is formed. On the other hand, if the grain boundaries are strengthened by precipitates and we have small grains which themselves are relatively weak, the latter deform easily giving low creep strength, but creep ductility is high, since intercrystalline cracks do not easily form. Many balanced alloys have a moderate mean grain diameter (well above 10μ), with grains and boundaries strengthened by dissolved elements and fine stable precipitates; over-aged large precipitates and brittle constituents at grain boundaries must be avoided.

Fortunately, a small chromium content improves the creep ductility of Mo and Mo-V steels, with little loss to their creep strength. 1% Cr-½% Mo steel and particularly 2¼% Cr-1% Mo steel have been much used for boiler tubes where moderate dimensional changes due to creep are tolerable, but long life without failure and therefore high creep ductility are necessary. About 1% Cr addition to Mo-V steel has been used for main steam pipes at 565° C where a good combination of properties is required.

2¼% Cr-Mo and low Cr content Mo-V steels have been used as cast components for the large turbine cylinders or casings. The horizontal flanges of the upper and lower half cylinders are bolted together using bolt and nut materials of adequate creep strength to maintain tightness. 1% Cr-1% Mo-0·8% V steel and high-strength nickel-base alloys have been used for bolts; these have thermal expansions similar to that of the low alloy steel castings, so that the bolt and flanges expand together on heating. The higher expansion austenitic steels are unsuitable in this respect as they would tend to lose tightness on heating.

Semi-stainless creep-resistant steels of about 8% chromium content, containing Mo, V and other elements, have been developed but not widely used in boilers or turbines. Stainless steels of about 0·1% carbon and 13% chromium content do not themselves have good creep properties, but again small contents of Mo, V and other elements confer considerable creep strength, with adequate creep ductility. Such alloys can be conventionally heat-treated to give high yield strengths and are used for blades at various temperatures throughout the H.P., I.P. and L.P. turbines; the natural chromic oxide (Cr_2O_3) surface layer, which occurs on stainless steels, minimises corrosion.

Where very good corrosion resistance, creep strength and weldability are required in a steel for use at the highest temperatures, e.g. for superheater tubes and steam pipes, an austenitic steel such as 16% Cr-10% Ni-2% Mo can be used. The high nickel content suppresses the normal FCC to BCC transformation, so the steel cannot harden and become brittle by this mechanism when rapid cooling occurs after welding.

In the H.P. and I.P. turbines, blades of some inches in length experience the highest temperatures, and steels or nickel alloys of high creep resistance must be used. At the end of the L.P. turbine, temperature is low, but the blades may be 3–4 ft long and must possess an exceptionally high yield point to withstand the rotational forces; the trailing edges of such blades also tend to be eroded by the impingement of condensed water droplets – a problem which is largely overcome by joining to the blade body a strip of hard cobalt-chromium alloy at the relevant part of the blade's profile. Another problem with all turbine blades is the avoidance of high vibrational stresses which might produce low-strain fatigue failure.

The large H.P. and I.P. turbine rotor shafts must operate with little creep strain for long periods at up to 550° C. Their yield and creep strength throughout their thickness must be adequate to support thermal stresses and also rotational stresses, particularly at the centre and at the blade fixing. The steel giving the best microstructure, with adequate properties from surface to centre of the rotor, has been found to be a 1% Cr-0·75% Mo-0·3% V steel of about 0·25% carbon content. Quenched and tempered rotor forgings of this composition give fine V_4C_3 precipitates throughout the steel thickness and a good balance of mechanical properties.

For the lower temperature L.P. turbine rotor and generator rotor forgings, creep is not a problem but the steels must be sufficiently alloyed to obtain adequate hardening throughout their thickness. Small Ni, Cr and Mo contents have been most common. A generator rotor, which may be 50 ft long, 4 ft in diameter and weigh 100 tons, must also be a steel of good magnetic properties if it is to convert mechanical into electrical energy efficiently. The design stresses on rotating L.P. and generator rotors are highest near the central axis which is usually bored with a hole of several inches diameter. L.P. rotor blades are frequently attached to separate discs or wheels which have been shrunk onto the cylindrical rotor shaft. The central perimeter of such wheels is subjected to very high stresses in service, the minimum yield strength required having increased from 80 000 psi to 120 000 psi in recent years as turbines have become larger. A low-alloy steel, suitably heat-treated, which meets the latest demands is a $3\frac{1}{2}$% Ni-$1\frac{1}{2}$% Cr-$\frac{1}{2}$% Mo steel, sometimes containing about 0·2% V.

An important property of generator and L.P. rotor forgings and wheels is their brittle/ductile transition temperature which should be below the minimum working temperature. Transition temperature is lowered by tempering a steel to lower strength, but for a steel of a given strength much can be done to lower the transition temperature by means of composition, heat treatment and small grain size. Some years ago rotor steels with brittle/ductile transitions of 100° C were common, but this has been reduced to 20° or 30° C for the $3\frac{1}{2}$% Ni-Cr-Mo steel.

The daily starting of power plant to meet peak electricity loads results in considerable mechanical stresses in turbine rotors and cylinders; a temporary high creep rate or yielding may occur on heating, while a reverse stress may appear on

METALS IN GAS AND STEAM TURBINES

cooling. In such conditions, high-strain fatigue failure must be guarded against, particularly in the cast cylinder.

Because brittle fracture and high-strain fatigue are assisted by the presence of any cracks or inclusions, and sometimes by porosity, non-destructive test methods are used to examine the soundness of large components during manufacture. A visual examination of a surface for cracks sometimes utilises a penetrating liquid containing a dye, or the application of a local magnetic field to the metal in conjunction with a fluorescent magnetic powder. Any faulty metal can often be removed and replaced by weld metal.

Internal cracks or porosity can be revealed by X-ray or gamma-ray photographs, or by ultrasonic waves which, in the case of large rotors, can be injected into the surface by a vibrating crystal probe, reflected back from the bore, and

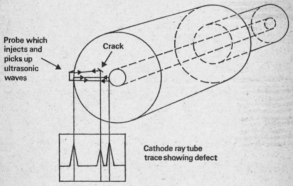

Fig. 17. *Ultrasonic testing of rotor forging.*

picked up by a second (or the same) probe at the surface; any crack or other defect reflects back some of the acoustic waves and the internal 'picture' can be displayed on a cathode-ray tube screen in the manner of a radar echo (see Fig. 17). It may take some hours to scan a large rotor in this way.

In steel-making practice some years ago, hydrogen which was picked up by the molten steel did not have time to escape before large forgings were finally cooled. This led to a problem of 'hair-line cracks' in rotors due to the hydrogen aggregating at local points of weakness. The ultrasonic flaw detector was able to reveal when such cracks occurred and heat treatments were subsequently applied to forgings which permitted the hydrogen to escape. The present approach to the removal of hydrogen and other impurities is to transfer the steel, usually from a basic electric arc furnace, into the ladle or mould under a moderate vacuum. This process is used for forgings for modern 500- and 660-MW turbine generators; these are so large that it might prove convenient to make two or more separate ingots, partially forge them, then weld them together before finally forging to size. This would be a suitable application for electroslag welding which can give a large high quality weld quite economically.

8. Materials in Nuclear Reactors

The particular requirements of nuclear reactors led to developments in some unusual metals – uranium, zirconium and beryllium, which had previously received little attention; the same can be said of some ceramics, notably uranium oxide and carbide, silicon carbide and graphite.

In a nuclear reactor a uranium atom can split or fission into two lighter atoms when bombarded by neutrons – minute particles contained in the nucleus of atoms; in splitting, the uranium releases spare energy in the form of heat, as well as other neutrons which can be used to produce fission in other uranium atoms (Fig. 18). This gives rise to a chain reaction releasing immense energy, some of which appears as penetrating gamma-rays and high-speed fission fragments of a whole range of elements, including gases such as xenon and krypton. Unfortunately, many of the fission elements produced are dangerously radioactive, giving off further gamma-rays and electrons before reaching stability.

The main function of the nuclear reactor is to control the fission process and provide heat in a form which can be used to generate electricity, usually by raising high-temperature steam for a steam turbine. The 'fuel' of uranium metal or its oxide or carbide is contained in a protective 'can' over which a high-pressure coolant is pumped to remove the heat. In commercial British gas-cooled reactors, the coolant is carbon dioxide (CO_2) and columns of 'fuel elements' are contained between a large volume of graphite (carbon) blocks; the whole reactor core is contained in a thick steel or pre-stressed concrete pressure vessel surrounded by protective shielding. The graphite's function is as a 'moderator' to slow down the neutrons, emitted by the fission process, to speeds suitable for fissioning further uranium atoms. The neutrons leave a trail of damage or defects as their energy is reduced on passing through the graphite.

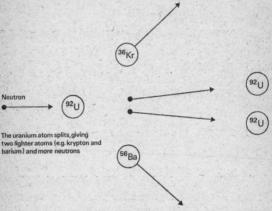

Fig. 18. *Fission of uranium.*

All atoms to some extent capture or absorb neutrons without fissioning, but carbon absorbs very few. Low neutron absorption is the common property of other materials used in such reactors, since neutrons are required for fissioning uranium and must not be lost by absorption. The number of neutrons present at any time in a reactor is controlled to give the required rate of fission and therefore of heat output and temperature; for this purpose 'control rods', containing elements such as boron which strongly absorb neutrons, can be raised or lowered an appropriate distance into the reactor core.

In early reactor experiments with uranium metal, the fission gases produced in the metal were found to present a problem when quite a small proportion of uranium atoms had been split. The gas atoms were able to diffuse to form high pressure bubbles which produced creep and swelling in the uranium. The uranium rods in a large reactor would reach temperatures of 500–600° C, and it was feared that swelling of the rods in service might fracture their metal cans and allow radioactive products into the carbon dioxide coolant; this would seriously contaminate the heat exchangers and blowers. It turned out that a correctly heat-treated uranium, containing very small amounts of alloying elements (e.g. Fe and Al), was able to minimise swelling; a fairly homogeneous distribution of fission product gas was produced without large bubble formation.

Another potential problem was uranium growth, a process whereby a polycrystalline bar of metal, whose grains were similarly oriented due to the mechanical forming process, became longer and thinner as it was irradiated with neutrons. The mobile uranium atoms appeared to favour re-attaching themselves to the particular crystal planes, which resulted in growth. This problem was mitigated by changes in the mechanical forming and heat-treatment processes, so that small randomly oriented grains were produced which minimised growth. The heat treatment consisted of quenching the uranium from its beta-phase (stable at about 700° C) to give a recrystallised alpha-phase, stable at lower temperatures.

Many different properties were required in the fuel cans – the externally finned tubes about 40 inches long and 1 inch diameter, sealed at each end by a welded end-cap. The canning material should have low neutron absorption (which rules out all but a few elements), compatibility with uranium at high temperature (i.e. not forming a brittle compound), corrosion resistance at high temperature to the high pressure CO_2 coolant containing a small moisture content, good weldability, freedom from ignition if overheated, adequate mechanical and creep strength, high creep ductility when the can is slowly strained due to fuel distortion or growth, adequate fatigue strength to combat any vibrational stressing due to the gas flow, and retention of a fine-grained (ductile) microstructure under strain and temperature conditions of operation. Such properties were required under conditions of constant neutron irradiation which introduced defects into the material's crystal structure.

Aluminium had been used to can uranium in early reactors,

but a thin barrier interlayer, for example of graphite, was necessary to prevent the formation of the compound UAl₃. Magnesium, which has very low neutron absorption, was compatible with uranium and a much better prospect for commercial reactors, if it could be alloyed to give adequate properties up to 450–500° C – exceptionally high service temperatures for a metal which melts at 650° C. After much research, an alloy with optimum properties, known as Magnox AL80, was developed containing 0·8% Al and 0·01% Be, and successfully used in all large British reactors employing uranium metal fuel.

Attempts were made to improve on Magnox AL80 because it exhibited significant grain growth at 450° C and only moderate creep ductility at 200–300° C. A magnesium-0·6% zirconium alloy (known as ZA) was the subject of much experimental work; this had a very fine and stable grain size and high creep ductility under all circumstances. ZA appeared adequate in all other respects, although its creep strength was slightly inferior to Magnox AL80; this was not of great importance since only small creep strength was necessary to prevent the can fins deforming. When tested in a reactor, however, it was found that the radioactive element plutonium, formed when a uranium atom absorbs a neutron without fissioning, diffused through the can wall and ultimately imparted radioactivity to the coolant; this phenomenon had not occurred with Magnox AL80 cans, possibly because their small aluminium content combined with the plutonium and stopped further diffusion. ZA cans were accordingly not used in British commercial reactors.

Magnesium alloys were also required for use as end-caps and other structural members where good creep strength was necessary, but where ductility was not so important as in the can. Manganese was found to be the principal element to impart high temperature creep strength. A small addition of Mn completely dissolved in magnesium at 575° C (see Fig. 19), and the alloy could be strengthed by slowly cooling or by ageing at a lower temperature when Mn precipitates formed within grains and at grain boundaries. A fairly large grain size and a Mn content of ½–1½% gave alloys that were satisfactory for use up to 450° C for the relatively low stresses involved.

Other magnesium alloys with small Zr and Mn contents were developed which combined good creep strength and ductility. About ½% Zr ensured nucleation of many grains giving a fine-grain size and good ductility, while about 0·1% or even less Mn, which has very low solubility in magnesium in the presence of zirconium, contributed greatly to creep

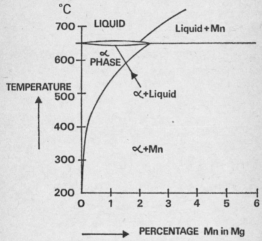

Fig. 19. *Magnesium-manganese phase diagram. 1½% Mn fully dissolves in Mg at 575°C, but this solution (the α phase) breaks down at lower temperature, giving Mn precipitates in nearly pure Mg.*

strength. Analysis of the alloy by the electron probe microanalyser showed that both Mn and Zr appeared in fine precipitates within grains and at grain boundaries. This alloy, which is ZA with a small manganese content, has been used in French nuclear reactors to can uranium; a thin interlayer of graphite or molybdenum between can and fuel was found to be effective in preventing plutonium diffusion.

It was also found that the ZA type of alloy could be improved in creep strength by holding at 600° C. This was believed to be due partly to grain growth and partly to the formation of zirconium hydride (ZrH_2) precipitates; a small amount of hydrogen is normally present in solution in magnesium alloys and is also available from moisture in the furnace atmosphere. Highest creep strength of all in magnesium-base material was obtained by recompacting Mg alloy particles which were slightly oxidised; when extruded to form a dense material, small dispersed particles of MgO behaved as very stable precipitates. Such alloys, however, had low creep ductility.

Zirconium has low neutron absorption and is potentially much stronger than magnesium at 450–500° C; unfortunately, it slowly corrodes in moist CO_2 to form the oxide (ZrO_2), and absorbs hydrogen to form the hydride (ZrH_2) which may embrittle the alloy. Research on Zr alloys showed that $\frac{1}{2}$–1% copper reduced corrosion significantly due to the formation of an adherent surface layer of mixed copper and zirconium oxides; it was also found that $\frac{1}{2}$–$1\frac{1}{2}$% molybdenum, or a small amount of tungsten or niobium, conferred high creep strength on such alloys after suitable heat treatment. A particular Zr-Cu-Mo alloy proved to be of considerable practical importance in the Magnox type reactors.

Zirconium has been used more extensively in a different type of reactor, common in the U.S. and Canada, which uses water as the coolant and delivers steam directly to a steam turbine. In Britain a prototype Steam Generating Heavy Water (SGHW) reactor has recently been built. The SGHW fuel element consists of a ceramic uranium dioxide (UO_2) fuel canned in Zr alloy tubes over which the coolant flows, reaching nearly 300° C as a mixture of high-pressure superheated water and steam. The fuel elements and coolant are contained in thick Zr alloy pressure tubes, so avoiding the large pressure vessel used for gas-cooled reactors. The moderator is 'heavy water', which is water (H_2O) in which each hydrogen atom contains an extra neutron in its positively charged nucleus, making it twice as heavy. Ordinary water contains a small proportion of heavy water, which can be extracted from it; heavy hydrogen, unlike ordinary hydrogen, is a small absorber of neutrons.

In heavy water moderated reactors the fuel element cans and pressure tubes have utilised a zirconium alloy (Zircaloy) containing about $1\frac{1}{2}\%$ of tin with small amounts of iron, nickel and chromium. Zircaloy has good corrosion resistance up to about 300° C due to the formation of a fairly impermeable oxide coating, but there has been concern about hydrogen absorption giving rise to embrittlement, particularly in the highly stressed pressure tubes; the absorbed hydrogen again appears as ZrH_2, often as needles at grain boundaries, so that deformation could give rise to cracking. Further research, however, has shown that this does not give rise to a practical problem.

Other Zr alloys of interest have been Zr-Cu-Nb and Zr-$2\frac{1}{2}\%$ Nb; the latter can be solution-annealed, quenched and aged so that Nb precipitates impart great strength to the alloy. Like steel and uranium, many dilute zirconium alloys have a crystal structure change at about 700–800° C – in this case from a BCC beta-phase to a CPH alpha-phase on cooling. Heat treatment accordingly produces complex microstructural changes; work-hardening of Zr–$2\frac{1}{2}\%$ Nb in addition to heat treatment gives strength values up to 100 000 psi. At high temperature this alloy's mechanical and corrosion properties are also good and it is a strong contender as a pressure-tube material.

Some attempts have been made to develop zirconium alloys suitable for use above 300° C, so that a water reactor could provide steam at a higher and more useful temperature. This was not successful by conventional alloying, but it may be possible to achieve much better strength at high temperature by utilising a dispersion of stable fine ceramic particles. Zirconium's high temperature strength is very disappointing in view of its high melting point (1852° C). Furthermore, like all metals used in reactors, its alloys are sometimes adversely affected by neutron irradiation damage.

An important recent development in reactor technology is the building of the first large, advanced gas-cooled reactor (AGR) power stations; these will produce electricity more cheaply than from coal- or oil-fired stations. The AGR is a development of the Magnox type of reactor and employs a graphite moderator and carbon dioxide coolant. As fuel it uses ceramic pellets of uranium dioxide, canned in thin stainless steel tubes; the latter reach temperatures above 700° C and enable steam to be raised at 565° C for supply to a modern steam turbine. The CO_2 coolant, which reaches 675° C in this reactor, has its minor 'impurities' closely controlled to minimise chemical reaction with the graphite.

One of the UO_2 fuel problems encountered was the possible escape of fission product gases from the UO_2 polycrystal into the space between the fuel and can. Although UO_2 can be made 100% dense, normal cold-pressed and sintered material contains several percent by volume of porosity, distributed as small closed pores in the ceramic. During fission the temperature at the centre of a half-inch diameter pellet can reach 1600° C, although its surface is only at about 800° C. It is possible for fission gases to diffuse to the pores and along grain boundaries to the pellet surface, where they would reduce heat transfer from fuel to can and so cause overheating. This is not a serious problem in normal practice, but if higher temperatures were used to increase efficiency, grain growth in the UO_2 would occur and fission gases might escape to the surface more easily. One method recently found of drastically reducing grain growth, at up to 2000° C in out-of-pile tests, was to add a small amount of finely divided iridium metal (Ir) or of another oxide, yttria (Y_2O_3). Perhaps stable iridium particles anchored grain boundaries in UO_2 by forming barriers to movement. The action of Y_2O_3 was probably more complex. Yttrium has a valency of three and combines with three O atoms for every two Y atoms, while the valency of U is four and UO_2 has four O atoms for every two U atoms. So when Y_2O_3 dissolves in UO_2, which normally has some vacant lattice sites for both U and O ions, there will tend to be a relatively large number of U sites filled by Y ions, but not enough O ions to sufficiently fill the anion sites available in the UO_2 fluorite lattice. The resulting relatively small number of U vacancies would make U (and Y) diffusion more difficult. As U in any case diffuses more slowly than O, and diffusion of *both* ions is required for grain growth, this would explain the effectiveness of yttria in reducing grain growth in UO_2. It is possible that fission gas release from the fuel would also be reduced.

Originally it was thought that the metal beryllium (Be), which has very low neutron absorption, might be developed with adequate ductility and corrosion resistance for canning

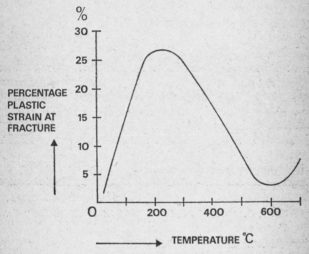

Fig. 20. *Effect of temperature on creep ductility of beryllium. Beryllium specimens were strained at 0·1% increase in length per hour. Each test was carried out at a constant temperature.*

the UO_2 fuel for AGR. Beryllium has a CPH structure, but at room temperature the crystal slips on only one set of planes (the base of the hexagon) and polycrystals are brittle. The metal becomes more ductile at higher temperature when other planes at different angles to the base also slip. It was found that the ductility could be much improved if minor metallic and oxide impurities were reduced, but further research showed that creep ductility fell to a low value at about 600° C (Fig. 20)

– a dangerous property in a canning material subjected to thermal cycles and creep strain. Beryllium also did not possess good corrosion resistance at 700° C in CO_2 and the metal was abandoned as a canning material.

One drawback had been that a powder fabrication route for preparing beryllium, involving cold pressing and sintering, gave low purity material. An arc-melting process, as used for other reactive metals, gave a higher quality more ductile metal with capacity for alloying to improve corrosion resistance; but the final choice for AGR was stainless steel which, although cheaper than beryllium, had higher neutron absorption and entailed consequent extra fuel expense.

A favoured austenitic stainless canning steel contained 20% Cr, 25% Ni and about $\frac{1}{2}$% Nb. Thin tubes of this material had adequate ductility and corrosion resistance up to 700–800° C, although vacuum melting and a very low carbon content were necessary to avoid large particle inclusions which might weaken a thin can wall. Minor impurities had also to be closely controlled in this steel.

Further development of gas-cooled reactors for commercial electricity generation is likely to be in terms of a highly rated uranium oxide or carbide fuel with a ceramic 'can'. The 'Dragon' high-temperature gas-cooled (HTGC) reactor experiment (a joint European venture in Britain) utilises an inert helium gas coolant and a fuel element of small UO_2 spheres, which are coated with pyrolytic graphite/silicon carbide layers and embedded in a graphite container. The helium could be used to raise steam for a steam turbine or be fed directly into an efficient high-temperature gas turbine before returning to the reactor. An alternative possibility is to develop the AGR reactor, retaining carbon dioxide as coolant, but using a silicon carbide outer fuel can which should resist corrosion to a higher temperature. High-quality silicon carbide and graphite are still the subject of much research.

A large HTGC reactor plant could probably produce electricity for 0·35–0·4d per unit, but the newer 'fast reactor', a prototype for which is being built, is likely to be developed to reach a figure as low as 0·3d per unit. This reactor utilises fast neutrons to produce fissions in a relatively small core of oxide fuel with no moderator. Heat is produced at a very high rate and removed by a liquid sodium coolant which is pumped between the core and heat exchanger, where high-pressure steam is raised at 565° C for a steam turbine. The fast reactor core is surrounded by a 'blanket' in which uranium atoms absorb neutrons without fission to form plutonium, itself a highly fissionable element. The plutonium oxide (PuO_2) is used to enrich the UO_2 fuel for use in the reactor core. By virtue of this process the reactor effectively produces more fissionable fuel than it consumes.

There are many material and component problems in the fast reactor due to the very reactive metal coolant and high thermal rating of the core. Sodium is compatible with stainless steel which can be used as a canning alloy, but oxygen must be meticulously kept from the sodium if Na_2O deposits are to be avoided in the coolant circuit. Some novel, as well as the usual corrosion and mechancial property problems also arise and will doubtless be solved.

9. Ceramics in Electronics and Power Equipment

The special electrical properties of certain elements and compounds place them in a unique category of 'semiconductors'. An important class of semiconductors includes the quadrivalent elements silicon (Si) and germanium (Ge), and the compound gallium arsenide (GaAs) – a 3:5 compound in which the average valency of the atoms is again four. Such ceramics form the basis of solid-state electronics and are used to make transistors, rectifiers and other components which replace the earlier vacuum or gas-filled valves used for amplification, rectification and other functions in electronic circuits.

Germanium was the first widely used high-quality semiconductor, but this has now been replaced in importance by its sister element silicon. The four valence electrons of the silicon atom are used for covalent bonding with its four neighbours, so that no electrons would appear to be available for electrical conduction. In fact, thermal activation processes in the silicon crystal results in electrons being dislodged from their normal shell to a slightly higher energy shell where they can behave as metallic conduction electrons. This is an 'intrinsic' effect which occurs in semiconductors; their resulting electrical resistance lies midway between that of metals and good insulators.

Another type of semiconduction, due to impurities, may occur in materials like silicon and is referred to as 'extrinsic'. If a small amount of phosphorus (valency 5) is added to a silicon crystal, its atoms will occupy lattice positions similar to silicon atoms, but they will each have an extra electron beyond the four required for covalent bonding in the crystal. The spare electrons appear in a higher energy shell where they are free to conduct; as they are negatively charged, the material is called an n-type semiconductor. It is easy to envisage that if trivalent impurity atoms, such as aluminium, are used to 'dope' silicon, there will be a deficit of bonding electrons for the crystal; the absent electrons can be regarded as 'holes' into which other electrons can quickly move, themselves leaving holes with the the absence of an associated negative charge. Such holes may

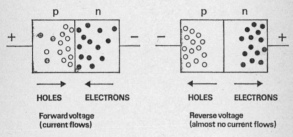

Fig. 21. *p–n Junction rectifier.*

be regarded as moving about like electrons, but acting as though they had a positive charge. This kind of semiconductor is referred to as p-type.

When pieces of p-type and n-type silicon crystal are in intimate contact and an electrical voltage is applied to electrodes connected to each side of the p–n junction, one of two things may happen (Fig. 21). If the positive electrode is on the p side, it will repel the positive holes towards the p–n junction, while the negative electrode on the n side will repel the negative electrons; at or near the junction, the electrons and holes will combine (cancel each other out) and a current will continue to flow. But if the electrode voltage is reversed, the holes and electrons are respectively attracted to their nearest electrode; the p–n junction is then polarised and no current flows. Such a 'device' is a rectifier or diode; if an alternating voltage is applied, only direct current flows – as electrons from the n to p side. There is, however, a limit to the 'reverse voltage' a rectifier can block, and much research has gone into raising this to several kilovolts for a single junction.

A transistor consists of two p–n junctions 'back to back' so that we have an n–p–n or p–n–p device. In the latter (Fig. 22), a forward voltage between the first p (emitter) region and the n (base) region drives the holes to the junction. If the base region is narrow, the holes cross through without combining with electrons there, and arrive at the second p (collector) region. The base/collector junction has a reverse fixed voltage

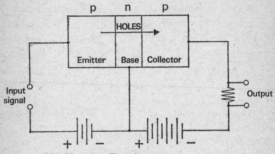

Fig. 22. *p–n–p Transistor*.

applied to it, so the holes which arrive form a current in the collector which is directly related to the number of holes and so the emitter/base voltage. The device can thereby amplify small voltage signals.

The purity of semiconductor silicon and its dopant concentration are very critical, and it is necessary to employ single crystals so that grain boundary irregularities are avoided. High-quality silicon (melting point 1410° C) is made from silane (SiH_4) or silicon tetrachloride ($SiCl_4$) and a high-purity single crystal can be prepared by the 'floating zone' technique – a special form of zone refining (see Fig. 23); this involves vacuum induction melting a narrow zone at one end of a heated silicon rod, and slowly moving the molten zone along the rod, allowing the silicon to solidify again in the wake of the zone. If the rod is held vertically, the 'floating zone' is kept in position by surface forces and there is no crucible to contaminate the silicon. The molten zone refines the silicon by preferentially retaining certain impurities in solution and carrying them along to the end of the rod; furthermore, the rod can be made into a single crystal by allowing the silicon to slowly solidify around a small 'seed' crystal at one end; the crystal then grows by the addition of atoms from the melt without further crystals

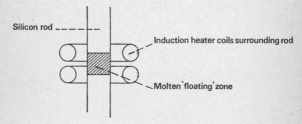

Fig. 23. *Floating-zone refining of silicon.*

nucleating. Some single crystal semiconductors are prepared by the Czochralski technique in which a crucible is used as a container and the single crystal 'pulled' from the melt by a seed crystal on the end of a slowly rotating vertical rod. Very low pulling speeds are necessary for crystals of high perfection.

Advantages of semiconductor devices are their small size and the low voltages required with little energy lost as heat. When a silicon crystal is sufficiently pure after repeated zone-refining (perhaps only one impurity atom in 10^9) it is cut into thin slices and 'doped' often by exposing the slice for a given time and temperature in a gas containing the necessary element; the latter diffuses inwards a certain distance to produce p or n material. p–n junctions must be made within a single crystal and quite intricate procedures are used, involving several stages, different dopants, and the use of masks to ensure doping of the correct area. Another successful procedure is to deposit silicon and the dopant together, from the vapour phase, onto the silicon substrate which may be already doped by another element. This process (epitaxial deposition) extends the same crystal lattice of the substrate. In any event the doped slice containing one or more p–n junctions is finally diced into small platelets for device preparation.

Silicon devices are used in both light electronics and high power applications. An example of the latter is the use of silicon rectifiers to convert alternating current to direct current for the traction motors of modern British Rail electric locomotives. These utilise a 25-kV a.c. overhead supply which is reduced to about 500 V by the locomotive's transformers prior to rectification.

In the new field of microelectronics, miniaturisation of semiconductor devices has been coupled with the concept of integrated circuits. A small disc of silicon may form the basis of hundreds of diodes and transistors, with conductor, resistor and capacitor elements vapour deposited as layers or films. In other instances, the substrate is a very flat sheet of glass or glazed ceramic, with local surface irregularities limited to a few hundred Ångstrom units; conductors, resistors and capacitors are deposited, again by a masking technique, and minute-size semiconductor devices can then be attached. Such miniature devices find wide application in general electronics and computers. An important new compound semiconductor now being used is gallium arsenide (GaAs).

Semiconductors, like all conductors, possess 'thermoelectric' properties. If a length of conductor has a hot and a cold end, electrons are excited to higher energy levels at the hot end, and a flow of electrons occurs giving a voltage between the ends; the voltage varies with conductor material so that if hot and cold junctions are formed between two materials in series, a net voltage is generated. Junctions between p- and n-type semiconductors give a larger effect than in metals and may be conveniently used as electric power sources, since they have fairly good electrical but poor thermal conductivity between junctions. p- and n-type lead telluride (PbTe) or bismuth telluride (Bi_2Te_3) can be used in the form of polycrystals, carefully cast from the melt to give a homogeneous and strong microstructure. Such thermoelectric generators can be used as small power sources where mains electricity is unavailable but where heat can be supplied to an array of junctions. Gas-heated generators in the kilowatt range have been made, but their thermal efficiency is under 10% compared with over 40% for the normal turbine generator.

Certain semiconductors such as cadmium sulphide (CdS), a 2:6-compound, possess interesting photoconductive properties. Radiant energy such as sunlight falling on the material excites electrons to the next higher energy shell where they confer electrical conductivity; this effect is used in light meters. If the Sun's radiation falls on a p–n junction, say of silicon, the excited electrons and corresponding positive holes can lower their energy by piling up on opposite sides of the junction; this gives a voltage which can provide power in an external circuit and is the principle of the solar battery, used in space vehicles.

Many oxide ceramics can behave as semiconductors, either due to non-stoichiometry or by virtue of a dissolved 'impurity' of different valency (such as lithium in MnO) and because certain cations behave in a multivalent manner. In the case of monovalent Li^+ replacing some Mn^{2+} ions in polycrystalline MnO, some of the remaining Mn ions give up an extra electron becoming Mn^{3+} to preserve electrical charge neutrality; in doing so they create holes which confer electronic conductivity. The iron cations in the double oxide magnetite, $FeO.Fe_2O_3$, have valencies of two and three and, depending on the precise state of oxidation, the magnetite polycrystal may have an excess of Fe^{2+} or Fe^{3+}. Electrons can hop from Fe^{2+} to Fe^{3+} ions giving electrical conductivity. These effects in oxides are much smaller than in silicon and oxides are not used for p–n junction devices. Such oxide semiconductors are used as 'thermistors', which are stable materials of high electrical resistance and which, unlike metals, sharply decrease in resistance (increase in semiconductivity) with rise in temperature; thermistors are used to measure temperature and to compensate for temperature effects in electronic circuits.

The mixed oxide polycrystal $BaO.TiO_2$ or $BaTiO_3$, known as barium titanate, can be made semiconducting by replacing a small proportion of Ba^{2+} ions by higher valency ions such as

La^{3+} or Th^{4+}; some of the Ti^{4+} ions become Ti^{3+} ions to compensate, and so permit 'electron hopping' between the two Ti ions. In this material an unusual resistance/temperature effect occurs at about 120° C where resistance suddenly increases about a thousandfold. There is a small but important crystal structure change at 120° C, and above this temperature it is believed that a slight excess of oxygen atoms at the grain boundaries trap the excess electrons and so form electrically insulating layers. The sharp resistance increase with temperature shown by doped barium titanate has led to its use as a temperature sensor, e.g. for overheat protection of motors; addition of strontium titanate ($SrTiO_3$) or lead titanate ($PbTiO_3$) to the material will lower or raise the temperature at which this effect occurs.

The reason that grain boundaries of semiconducting barium titanate are not insulating below 120° C is related to its 'ferroelectricity' below this temperature (the 'Curie temperature'). Ferroelectric ceramics have a non-symmetrical crystal structure in that the cations and anions are positioned so that the net positive and negative electrical charges are displaced from each other, or polarised. The positive–negative 'dipoles' point in different directions in different 'domains' of each crystal. The total electrical field effect in a polycrystal usually cancels out, but the dipoles can be made to line up by cooling the ceramic through its Curie temperature in the presence of an electric field ('poling').

Single crystals of 'ferroelectrics', as well as some polycrystals, also exhibit useful 'piezoelectric' properties; when a mechanical pressure and corresponding elastic strain is applied to a poled piezoelectric ceramic disc the degree of polarisation changes, giving positive and negative charges on opposite faces of the disc. Equally, a voltage applied to a piezoelectric ceramic produces corresponding elastic strain; such ceramics are used as 'transducers' of electrical to mechanical energy and the reverse. A high-frequency alternating voltage applied to a 'piezoelectric' gives mechanical or acoustic vibrations which can be transmitted through a solid or liquid. Transducers can equally pick up sound or other acoustic vibrations and trans-

CERAMICS IN ELECTRONICS AND POWER EQUIPMENT 95

late them into electrical signals. They find applications in probes for non-destructive ultrasonic testing (Chapter 7), in submarine detection, microphones, pick-ups, high frequency electronic oscillators and many industrial fields. Barium titanate has good piezoelectric properties, but a favoured material is lead zirconate titanate (PZT).

Ordinary polycrystalline $BaTiO_3$, by virtue of its polarisation in an electric field, can store more electrical energy than most insulators. The electrical charge stored per volt by the

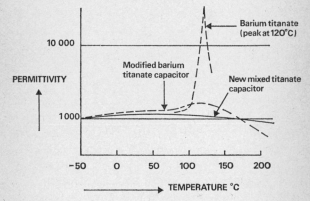

Fig. 24. *Permittivities of some titanate dielectrics.*
(*Courtesy of* SCIENCE JOURNAL.)

insulator or 'dielectric' between the metal electrodes of a capacitor, of standard geometry, is proportional to a quantity known as the permittivity. Common ceramics, glass and plastics have permittivities between about 3 and 10 (air=1), but barium titanate's permittivity is several hundred, rising to many thousand at its Curie point (120° C) where it loses its ferroelectric properties. In consequence, barium titanate is used to make high permittivity capacitors, some additions being made to the material to even out its permittivity over a region of temperature. One unusual dielectric has been developed consisting of a mixture of magnesium titanate with strontium bismuth lead titanate which has an almost constant permittivity of 1000 between $-50°$ C and 200° C (Fig. 24).

There are, however, two disadvantages of ferroelectric capacitors; in the first place, a piece of the dielectric cannot withstand such a high voltage as a more common ceramic or plastic; and, secondly, ferroelectrics show greater 'dielectric loss' under a.c. use, the lost energy appearing as heat.

Similar in some ways to ferroelectric ceramics are the polycrystalline ferrimagnetic ceramics. These usually have a crystal structure based on spinel ($MgO.Al_2O_3$), and contain domains which are polarised, not by electrical ionic charges, but magnetically, giving a north and south pole. As mentioned in Chapter 5, the magnetic polarisation of domains results from an imbalance of spin direction of electrons in sub-shells of the atoms. The direction of magnetisation of the small domains can be rotated and lined up by an external magnetic field to make a large volume of material into a magnet.

The most important magnetic ceramics, known as ferrites, have the general formula $XO.Fe_2O_3$ where X is a divalent atom such as Mg, Mn, Ni or Zn. Ferrites appear as typical brittle electrically insulating ceramics, a so-called 'soft' ferrite being one which is easily magnetised or demagnetised such as magnesium-manganese ferrite; these materials can be magnetised by means of an electric current and are used in computers as small wire-wound rings which, according to the direction of magnetisation they are given, store numerical information. Soft ferrites are also used as a form of switch in radar equipment for extremely high-frequency 'microwave' currents which interact with the ferrite. Because of their high resistance and low loss, ferrites are used as high-frequency transformer cores for radio and television, but they cannot be magnetised to the same extent as iron or iron-silicon alloy normally used in electric motors and generators.

The 'hard' ferrites, which are not demagnetised easily, are comparable to the complex metallic permanent magnets which often have a high cobalt and nickel content. Typical hard ferrites are based on barium ferrite ($BaO.6Fe_2O_3$) and are used in small electric motors to provide the fixed magnetic field required. Both ferrimagnetic and ferroelectric ceramics are made by cold compacting mixed oxide powders of adequate purity and sintering in a controlled atmosphere, often at 1200–1400° C.

In semiconducting ceramics, we have been concerned with conduction by electrons. The other important mechanism of electrical conduction is by the ions, which become more mobile as temperature is increased. If an electric field is then applied to the ceramic, cations will drift to the negative electrode and anions to the positive, resulting in a small current. Usually either the anion or cation is more mobile and can contribute more to the current.

Electrical energy can be stored as chemical energy and can be released by ion transport or electrolysis of the above kind. This is the basis of the battery. In many batteries ions flow between electrodes in a liquid or paste, but in certain modern lightweight batteries being developed, the 'electrolyte' is a ceramic. One such battery, which may be used for the electric car, employs an electrolyte based on the double oxide $Na_2O.11Al_2O_3$ (misnamed beta alumina), which separates electrodes of liquid sodium and sulphur. The special lattice structure of beta alumina enables the Na^+ cations to travel quite rapidly through it at a temperature as low as 300° C. Sodium atoms give up their electrons at the sodium electrode and enter the ceramic as Na^+ ions. The electrons provide an electric current in the external circuit to the sulphur electrode where they enable the formation of sodium sulphides (e.g. Na_2S_5) with the sodium ions emerging from the electrolyte. On recharging the battery, the externally applied voltage pulls back sodium ions to the sodium reservoir.

Ceramic electrolytes may also find use in 'fuel cells' which, unlike the battery, have their source of chemical energy supplied externally. One example is an electrolyte of zirconia (ZrO_2) containing about 15% calcia (CaO); this time the anion (O^{2-}) is the charge carrier, at about 1000° C. Pure ZrO_2 undergoes crystal structure changes on heating which tend to cause fracture, but the addition of several percent CaO, MgO or Y_2O_3 stabilises the fluorite structure; furthermore, the addition of cations of lower valency than zirconium produces an anion deficiency with consequent greater oxygen mobility in the lattice. In this cell, oxygen ions can travel through the zirconia from one special electrode to the other, where they give up their electrons to the external circuit and by combustion oxidise a fuel containing hydrogen and/or carbon; natural gas (mainly methane – CH_4) might be used. Electrons reaching the first electrode from the external circuit permit the oxygen supplied (from the air) to be ionised and enter the lattice.

Both batteries and fuel cells supply direct current and require a continual supply of ions; if such ions are not available and a d.c. voltage is applied to an ionically conducting ceramic, it soon polarises and no further ionic current flows. But ceramics based on thoria (ThO_2) and zirconia (melting points 3300° C and 2700° C respectively) are available as dense impervious materials which conduct both ionically and increasingly by electronic semiconduction from 1000° C to 2000° C and can be used as electric furnace elements; apart from certain oxides, no other material is sufficiently stable at 2000° C unless enclosed in a vacuum or inert medium.

Zirconia and thoria are not typical oxides in regard to electrical conductivity; alumina, silica, magnesia and beryllia (BeO), by contrast, are good electrical insulators up to high temperatures. In applications where both high-voltage insulation and good mechanical properties are required, as for sparking-plug bodies and in large electronic valves, ceramics with high alumina content (85–95%) are frequently used. The remaining content of such ceramics is a glassy phase containing silica.

Most common insulators for switchgear and other high-voltage equipment are made of porcelain; this consists of a glassy alkali alumino-silicate matrix containing crystals of quartz (SiO_2) and mullite ($3Al_2O_3.2SiO_2$). It is made relatively cheaply by firing at about 1300° C a moulded mixture of clay (e.g. kaolinite – $Al_2O_3.2SiO_2.2H_2O$), sand or flint (SiO_2) and an akali such as potash feldspar ($K_2O.Al_2O_3.6SiO_2$), the latter being a major constituent of a common igneous rock.

For overhead power lines, common soda-lime glass insulators are often used; these are toughened, like car windscreens, by rapidly cooling the surface layer of the glass so that it is put in compression when the inside contracts on cooling. This makes tensile fracture more difficult, but if it does occur, the glass shatters.

When a combination of very good insulation and mechanical properties is required, a modern alternative to the high alumina ceramic is a so-called glass-ceramic. Certain compositions of glass can be first formed to the shape required, then heat-treated so that they devitrify to give an almost wholly polycrystalline ceramic. By careful control of crystal nucleation and grain size, insulators can be made of three or four times the strength of porcelain or glass and equal to the best alumina ceramics.

Between glass or porcelain and the higher-quality alumina and special glass-ceramics, there are many more conventional ceramic insulators. Porcelain can be upgraded in strength by increasing its alumina content, while the addition of zircon ($ZrO_2.SiO_2$) both improves insulation and nearly doubles the strength. For best insulation the glass content should contain a minimum of the alkali ions found in porcelain. Ceramics based on mullite ($3Al_2O_3.2SiO_2$) in particular have good high-temperature properties; if, in addition, resistance to thermal shock and therefore a low thermal expansion is required, as in highly rated fuse bodies, ceramics based on cordierite ($2MgO.2Al_2O_3.5SiO_2$) are used. For high-frequency electronic applications where very low electrical loss combined with good mechanical strength is necessary, bodies based on forsterite ($2MgO.SiO_2$) are common; these have a higher thermal expansion matching that of several metals to which they can be joined. Ceramics based on steatite ($MgO.SiO_2$) also have good general properties and can be readily coated by a firmly bonded metal layer (metallised) for joining to a number of metals.

Pure alumina (melting point 2050° C) is of special interest in that, unlike the 85–95% alumina ceramics, it is completely free of glass and so can be used at temperatures up to about 1500° C or higher. Normally sintered pure alumina contains some enclosed porosity, but techniques have recently been developed for eliminating porosity and producing a fully dense and almost transparent ceramic. This can be done by making a small addition of another oxide (e.g. MgO) to the alumina powder; this controls grain growth and assists in the elimination of pores during sintering. Fully dense alumina shows very low electrical loss at high frequency and its mechanical strength can be as high as 50 000 or 100 000 psi, depending on grain size. One use for it is the windows of high-power microwave valves used in radar, where its high transparency to this radiation prevents the window from overheating – a trouble that can lead to runaway temperature in less transparent ceramics, which give rise to a significant power loss.

An important property of fully dense alumina is its resistance to corrosive attack by alkali metals at high temperature. Tubes of the material are now being used to enclose the high temperature sodium discharge in new lamps giving a high efficiency 'golden white' light output. Fully dense alumina would also be of interest for 'thermionic' electricity generators which would employ cesium (Cs) vapour, or for duct insulators for magneto-hydrodynamic (MHD) generators; in the latter, a high-temperature gas stream from a nuclear reactor or heat exchanger is made conducting by adding potassium ions, and passed between electrodes and the poles of a powerful magnet to generate electricity; the cooler emerging gas can then be utilised by a turbine generator. The overall efficiency of electricity generation from the MHD and turbine generator would be very high, but this has not yet been achieved.

High-density alumina is a fairly good thermal conductor and so finds use as a substrate for microelectronic circuitry, where the heat generated must be adequately conducted away. An even better, though more expensive thermal conductor, is high-density beryllia (BeO); this should find increasing use as a substrate material.

A new function of ceramics has arisen in the field of the LASER (Light Amplification by Stimulated Emission of Radiation). Single crystals of ruby (alumina with a small chromium content), calcium tungstate ($CaWO_4$) and some other oxide ceramics, doped with ions of the rare-earth metal neodymium (Nd^{3+}), can behave as electronic amplifiers of light waves. A high-intensity pencil of light is generated from the input energy in such a way that the output waves are in phase with each other, or 'coherent'. Very narrow high-power beams of a given wavelength can be transmitted, and received a great distance away, e.g. in outer space. The pencils of light can also be 'modulated' like high-frequency radio waves or microwaves, but, because of their smaller wavelengths can communicate far more information.

10. High Temperature and Space Technology

Many metals and ceramics are prepared and fabricated above 1000° C, but not many are used above this temperature. Several 'refractory metals' have useful strength between 1000° and 2000° C, but corrode by oxidation unless they are protected by a special coating or inert atmosphere. Some ceramics are more stable at high temperature; certain oxides can be used at 2000° C, while a number of silicides, carbides, nitrides and borides of elements have a useful combination of properties between 1000° and 2000° C.

Of the modern developments which involve materials at the extremes of temperature, that of rockets and space vehicles is perhaps the most interesting. The vast power required to launch a large rocket in the Earth's gravitational field, especially if a satellite is to reach a height of hundreds or thousands of miles where it may orbit the Earth, necessitates efficient fuel economy and the use of very high temperatures. The thrust of a first stage rocket motor is obtained by burning solid or liquid fuel and ejecting the combustion products through a nozzle. On the combustion chamber side of the nozzle throat very high pressure builds up, with a gas temperature possibly up to 3000° C; outside the profiled nozzle throat the gas expands and cools. The nozzle itself, which may be up to several feet in diameter, experiences extremely high temperature (perhaps above 2000° C), severe thermal shock and corrosion effects; in addition, any alumina particles, from solid fuel which contains aluminium, tend to erode the throat. Second- and third-stage rocket motors have an easier task because the vehicle is then lighter; liquid oxygen and hydrogen often forms the basis of combustion and temperatures are much lower. Fortunately, the required operational life of many components is quite short compared with those in other applications.

Much of the structure of rockets and their contents do not experience high temperature, and considerable use is made of light alloys of aluminium and magnesium, as in aircraft. An orbiting space capsule or satellite itself experiences a unique combination of circumstances. It is effectively in a vacuum where 'protective' oxides do not form on an abraded metal surface, so metals easily cold-weld or stick together tenaciously. External temperature falls in the absence of Sun to about $-270°$ C ($3°$ K), but part of the vehicle receives direct radiation from the Sun; besides heat, this includes powerful and possibly dangerous ultraviolet rays, X-rays and cosmic rays. The vehicle is also bombarded by a spectrum of small high-speed particles (meteorites) which fortunately have not proved to be too troublesome.

A considerable problem arises when a vehicle re-enters the Earth's atmosphere. The density of the atmosphere falls rapidly with height above the Earth until, at 30–40 miles up, there is very little air. If a missile travelling at 10 000 m.p.h. enters the air belt at a steep angle, it should reach the Earth in much less than a minute, but unless it is specially designed, frictional heating by the air would vaporise it at white heat; this happens in the case of certain meteoric bodies ('shooting stars'). To avoid very high temperature problems, the missile must be carefully profiled and its nose cone, in particular, made of material which can withstand and dissipate the heat. A manned space vehicle orbiting the Earth at about 20 000 m.p.h. could not tolerate such direct entry and must enter the atmosphere at an oblique angle, perhaps 5 degrees to the horizontal, and make a slower approach. This implies many minutes of descent in which considerable heat is generated at the vehicle surface. The required materials of construction are therefore good thermal insulators, heat-resistant solids and materials which can safely absorb a large amount of heat.

Space technology poses not only the normal materials

problems of strength, ductility, lightness, corrosion resistance and compatibility, but also unusual problems of very high temperature and thermal shock. The latter especially occur for combustion chambers, nozzles, nose cones and leading edges of re-entry vehicles which may exceed 1500° C. Before noting how such problems can be tackled, we shall examine more closely the nature and some other uses of the high temperature metals and ceramics which are available.

The high melting-point metals are in the B groups of the periodic table. The rather special six 'platinum metals' of Group VIIIB (Ru, Rh, Pd, Os, Ir, Pt) have melting points up to 2700° C; they are very dense and expensive, but the corrosion resistance, in particular of platinum, rhodium and iridium, has led to numerous industrial uses in chemical plant and as high-temperature thermocouples and crucibles.

The metals of greatest practical interest are in Groups IVB, VB and VIB, although here we will ignore the three radioactive elements (thorium, protoactinium and uranium) which do not have high melting points and are of very specialised interest. In Group IVB (Ti, Zr, Hf) only hafnium has a high melting point (above 2000° C) and can be considered a refractory metal; it is relatively rare and costly but has become more available in recent years as a by-product of nuclear reactor grade zirconium production; Hf is a fairly strong neutron absorber and so must be extracted from the zirconium with which it occurs. Hafnium has good oxidation resistance and when alloyed with niobium or tantalum can be considered for use well above 1000° C; its alloys, however, are not yet highly developed. Of the other two metals, zirconium is used mainly in nuclear reactors and does not have good high-temperature properties, while titanium is again confined to use at moderate temperatures (similar to those for steels), but is of growing interest and importance due to its corrosion resistance, lightness and high-strength capabilities. Titanium's future probably lies in the chemical and process industries and in machinery where light-reciprocating or rotating parts are required, and in the aerospace field. When Ti is alloyed with small percentages of aluminium, with such elements as manganese, molybdenum, vanadium or tin, tensile strengths of 150 000 psi can be achieved. Ti alloys are used for gas turbine compressor blades, ducts and casings; they are also strong contenders for future supersonic aircraft skins where temperature is too high for aluminium alloys.

The BCC metals of Groups VB (V, Nb, Ta) and VIB (Cr, Mo, W) have many similarities, although a notable difference is the relatively high brittle/ductile transition temperatures of the three Group VIB metals when even small amounts of impurities are present. Vanadium, like chromium, melts below 2000° C, but the metal has much poorer oxidation resistance than chromium. Vanadium, however, can be protectively coated with its silicide (VSi_2) and has the advantage of moderate density, though not as low as titanium; it can be alloyed, for example, with niobium and titanium, to give very high tensile strengths, but the use of vanadium is still very limited.

The four remaining, truly refractory metals (Nb, Ta, Mo, W) are more readily available and of considerable interest for use at temperatures up to 2000° C or (for tungsten) 3000° C. Niobium and molybdenum are not much denser than steel and melt at 2415° and 2610° C respectively; they form the basis of alloys which, with protective coatings, can be used up to 1300° C, or at low stress to 1700° C. Tantalum and tungsten are more than twice as dense as steel, but melt at 2996° and 3410° C respectively; tungsten is of some importance up to 2000° C if protected from oxidation, and can even support small stresses as lamp filaments at about 3000° C.

Niobium and tantalum are basically more oxidation resistant than molybdenum and tungsten, some alloys of niobium combining relative lightness with corrosion resistance for short periods at 1100–1200° C. Nb has been alloyed with such combinations as W, Mo and Zr or Ti and Zr to give useful strength at 1300° C or higher, and so is of interest for future gas turbine blades and aerospace components. To utilise the strength of Nb alloys at the highest temperatures, research on protective ceramic coatings has been very active and has resulted in the use of silicide mixtures; in an oxidising atmosphere these form a thin surface of glassy silica (SiO_2) which drastically reduces further oxidation. On niobium alloys a coating of $NbSi_2$ containing 10% of VSi_2 or $TiSi_2$ gives good protection to 1300° C, while on Ta, Mo and W alloys, similar silicide mixtures, often rich in the silicide of the parent metal, give similar protection. Sometimes a special glaze overlay is applied to the silicide coating so that any cracks produced by thermal stress on heating are quickly sealed up. Siliciding a surface is carried out either by high temperature vapour reaction with a silicon compound, or by a diffusion treatment in which the metal component is heated in intimate contact with a compact powder mix of appropriate composition ('pack diffusion'); the thermal expansion of such coatings can be matched to the substrate metal to minimise thermal shock cracking. Quite robust coatings are produced despite their inherent brittleness.

Alternative types of coating for refractory metals, based on the formation of an adherent protective oxide (BeO, Al_2O_3 or Cr_2O_3), are the beryllides, aluminides and other intermetallics, such as $NbBe_{13}$, NiAl, TiAl, and Cr_2Ti. These are useful only to about 1200° C, but the aluminides are interesting in that they possess a little ductility at low temperatures. Such coatings, as well as silicides, can often be applied at conveniently low temperatures by electrolysis of fused low melting point compounds ('metalliding'). Oxide coatings such as alumina or zirconia afford greatest protection of all, but are very brittle; some borides give protection up to about 1500° C, while a mixture of 10% $MoSi_2$ with ZrB_2 or Al_2O_3 can give useful protection up to 1700° C.

When alloys are required with better high-temperature properties than those of niobium, alloys of tantalum may be used. Ta alloys containing such additions as tungsten, hafnium, niobium and vanadium can be coated, for example with $MoSi_2$, and used as sheet in construction of combustion chambers for rockets; in such applications the alloys are not highly stressed and may be used up to 1600° C.

Molybdenum and tungsten are the most commonly used refractory metals, but both oxidise at quite moderate temperatures and must normally be protected by a suitable solid, gas or by vacuum. Pure Mo and its alloys, for example containing titanium, are fairly ductile and are available in fabricated forms such as sheet. They have many high-temperature applications, including components in rocket-nozzle construction. Mo is used in wire form for electric furnace elements (protected by hydrogen), where temperatures up to 1800° C are required for sintering oxides to high density. It is used in thicker sections for electrodes both in vacuum devices and to pass current into molten glass – to heat it to 1500° C in large melting furnaces. The metal also forms high-temperature compounds such as $MoSi_2$, itself usable for electric furnace elements without special protection.

Tungsten and its alloys, for example with molybdenum, have a similar range of applications and even more remarkable high-temperature properties than molybdenum and its alloys. Tungsten is best known as common lamp filament wire, where it has a life of many thousand hours at about 2400° C. If drawn wire of very pure tungsten were used as a lamp filament, it would recrystallise at under 2000° C and its grains rapidly grow to occupy the width of wire; grain boundaries lying across the wire thickness would then be a source of weakness, giving rapid grain-boundary creep and failure under the filament's own weight. To prevent this the tungsten powder is 'doped', prior to sintering, with small amounts of alkali and silica, some of which are retained in the final wire. On recrystallisation the additives modify grain growth so that boundaries exist as long irregular paths, parallel to the wire axis, ultimately reaching the wire surface. The polycrystal of long narrow grains then behaves more like a single crystal and creep cannot readily occur across the wire axis.

It has been found that when tungsten is alloyed with rhenium (Re), a fairly rare Group VIIB refractory metal (melting point 3180° C), the alloys are much more ductile than either metal itself; W-Re alloys have found use as filaments in lamps which must withstand vibration, and as thermocouple wires for measuring temperatures up to 2500° C.

Pure Mo and W can be prepared by reacting the impure metal powder with iodine (I) vapour at high temperature to form a pure gaseous metal iodide; this is then decomposed on a still hotter electrically heated wire of the metal. The deposited metal builds up on the wire to become a pure rod which can itself be mechanically worked to wire or some other form. This chemical 'trick' is utilised within a small fused quartz envelope in the 'quartz-iodine lamp' (Fig. 25) which enables the filament to run at about 3000° C, giving greater efficiency of light output. When running, the tungsten which evaporates from the filament combines at lower temperatures with iodine

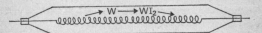

Fig. 25. *Quartz-iodine lamp. In practice, the simple tungsten-iodine cycle shown is modified by a small moisture content present, making the chemical reactions more complex.*

present in the inert gas (argon) of the lamp; at the same time the tungsten iodide present is decomposing and depositing tungsten on the hottest points of the filament, which are the points where it would otherwise 'burn out'. A more recent development, now called the tungsten halogen lamp, utilises an organic compound of bromine (Br), another of the Group VIIA 'halogen' elements, which can give better results than iodine.

Molybdenum and tungsten are normally consolidated by sintering the pressed metal powder, previously obtained from compounds produced from the purified ore. At the sintering temperature of tungsten, many of the impurities are volatilised leaving a fairly pure metal. Other metals can often be obtained in a fairly pure state by a melting process in vacuum, when many impurities again vaporise. The vacuum arc melting process, in particular, is used to produce titanium, zirconium, niobium, tantalum and beryllium. It is claimed by some that electron beam melting is a more controllable process for such metals and is capable of giving a finer grain-size ingot which is less likely to crack on subsequent mechanical working.

The high-temperature ceramics of most interest are oxides, but carbides, nitrides and borides, as well as carbon itself, are of growing importance. The carbides are the highest melting-point materials known, tantalum carbide (TaC) and hafnium carbide (HfC) melting at about 3900° C. Chromium carbide (Cr_3C) has a lower melting point, but has better than usual oxidation resistance and is light and wear resistant. The extreme hardness of tungsten carbide (WC) and titanium carbide (TiC), which are used as the basis of cutting tool tips, falls off above 1000° C which can be reached at the tool's working tip. Two non-metal carbides, boron carbide (B_4C) and silicon carbide (SiC), do retain good high-temperature hardness and abrasive properties and, in the form of dispersed particles, are used in grinding wheels; they can also be prepared as fully dense (pore-free) bodies. SiC enjoys the same type of oxidation resistance as silicides since it forms a glassy layer of silica (SiO_2) which gives protection for limited periods at up to 1600° or 1700° C.

Fully dense SiC has been prepared by decomposing the vapour of methyltrichorosilane (CH_3SiCl_3) at 1400° C on a heated graphite substrate. Tubes of such 'pyrolytic' SiC can have strength up to 100 000 psi and very good high-temperature stability and wear resistance. Fully dense pyrolytic graphite can also be prepared and has been deposited alternately in thin layers with silicon carbide, with the result that the composite has high thermal conductivity and good resistance to thermal shock. Such layers are used to coat uranium carbide or oxide fuel particles for high-temperature nuclear reactors. Carbides as nuclear fuels have greater thermal conductivity than oxides and much research has been taking place on high-purity dense carbides of uranium and plutonium.

Polycrystalline carbon or graphite will oxidise below 400° C, but it is an unusual material in that its mechanical strength increases up to 2400° C and is still relatively high at 3000° C. Since graphite has low density and good thermal shock resistance, it is attractive for use in rocket components such as nozzles. For moderate temperature use in an oxidising atmosphere, graphite can be coated with a layer of silicon carbide.

Ordinary graphite, which is porous and rather weak

mechanically, is made by firing blocks of carbon particles pressed with pitch as a binder; volatile constituents are driven off and at nearly 3000° C most of the carbon atoms form minute graphite crystals. Different degrees of graphitisation occur according to details of the process and a range of densities and mechanical and electrical properties can be obtained. Inorganic additions can be made, for instance to improve the corrosion behaviour of graphite electrodes. Vast quantities of fairly pure graphite blocks have been used in recent years as nuclear reactor moderators. For future ceramic fuel reactors, denser impervious graphites will be required; one method of obtaining higher density is to impregnate the porous graphite with organic material and refire it.

Nitrides of B-group metals, like carbides, are electrical conductors, but the nitrides of aluminium, boron and silicon (AlN, BN, and Si_3N_4) are good electrical insulators, the last two of increasing engineering importance. BN and Si_3N_4 can be consolidated from powder or deposited pyrolitically in dense form with strengths up to 100 000 psi. BN is a white somewhat resilient solid, resembling graphite in crystal structure and properties, but very inert and oxidation resistant up to 1500° C. It is interesting that a very hard cubic crystal form of both BN and carbon (diamond) can be made artificially at high temperature and extremely high pressure. Ordinary hexagonal BN is a good high-temperature dielectic and is transparent to infra-red and microwave radiation up to 1300° C. It has good thermal shock resistance and is used for crucibles for reactive metals and for engine and space vehicle components. Some corrosion protection can be afforded by firing a BN powder coating on a metal surface, although such a layer is inevitably porous.

Shapes of silicon nitride are now made by part-nitriding a porous sintered silicon body, machining to size in this condition and finally completing the nitriding at 1400° C, when hardly any further shrinkage occurs. Such material contains pores and has a strength of about 30 000 psi, but can be used in air up to 1200° C and has a useful combination of properties; it has good thermal shock resistance due to its low thermal expansion, resists attack by many reactive metals and acids, and even slightly increases its strength up to 1200° C. Again, it has been used as moving or stationary parts in piston engines, turbines and rockets; composite Si_3N_4/SiC mixtures have been made for corrosive and arduous high-temperature use in high-duty spark plugs.

The borides of B-group metals form another class of refractory compounds, this time having good electrical conductivity, greater hardness than that of carbides, and oxidation resistance in air up to at least 1200° C. Zirconium and titanium diborides (ZrB_2 and TiB_2) melt at about 3000° C; they are extremely wear-resistant and their electrical conductivity is superior to that of many metals. Possible uses are as heavy-duty electrical contacts, and as electrodes in liquid metals or in the corrosive gases of MHD electricity generators. Borides are also being considered as a more wear-resistant alternative to carbides for cutting tool tips, but this research has not yet produced materials for general use.

Boride powders, like many refractory ceramics, are difficult to sinter, and it is common to use hot pressing if a large amount of porosity is to be avoided. A typical arrangement employs a furnace tube which encloses a cylinder with single or double pistons moved hydraulically to compress the body (Fig. 26). If metal dies such as strong nickel alloys are used, temperature is limited to about 1000° C. This is sufficient to achieve a high density in many materials, but refractory dies such as graphite may be used in an inert atmosphere to press small pellets at over 2000° C; pressures used are often 20 000–30 000 psi. Compared with normal sintering of cold-pressed powders, hot pressing gives a higher density for a given temperature, or enables a lower temperature to be used. Besides borides, self-bonded and dense nitrides and carbides can be made by this process; sometimes a metal or more plastic phase is added and it is also possible to promote a chemical reaction to form a compound during hot pressing.

A newer and more sophisticated technique is isostatic hot pressing, known also as gas-pressure bonding. Here the powder is first cold compacted and perhaps pre-sintered to avoid fragility; it is then sealed in an evacuated container of thin metal; this is put into a furnace in a pressure vessel and a high pressure of argon or helium applied, perhaps 15 000 psi or more at up to 1500° C. This eliminates the die problem and can uniformly consolidate bodies of several inches diameter

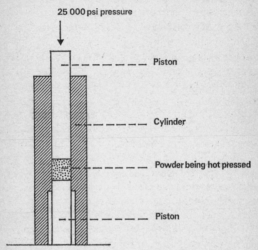

Fig. 26. *Hot pressing of powders. Heat may be supplied by means of a furnace or by passing a large electric current through the assembly.*

and length. The equipment can also be used for combining sintering with a chemical reaction and for bonding metals to ceramics. It opens up many possibilities, but perhaps its main advantage is in preparing uniformly and densely sintered bodies at lower than usual temperature; this reduces unwanted effects of heating such as grain growth or diffusion which may impair properties.

'Composite materials' is a term embracing the combination in various forms of different metals and ceramics in order to obtain better properties than is possible with individual ma terials. Tungsten carbide sintered with cobalt was one of the earliest ceramic-metal, or 'cermet', materials which attempted to combine the carbide's hardness with the metal's toughness; this formed the basis of an industry producing cutting tools, dies, rock drills and stone crushers; many such 'cemented carbides' have bend strengths exceeding 200 000 psi. A competitive cutting tip with extreme wear resistance is fine-grained alumina, but its strength cannot much exceed 100 000 psi and it remains brittle. One cermet composite now available is a 50/50 steel/TiC mixture – half-way between conventional cemented carbide and highly alloyed tool steel.

Many cermets, containing sufficient metal to impart some toughness, have been developed with useful strength and oxidation resistance up to about 1200° C; examples are TiC/Ni-Cr and Al_2O_3/Cr. A main problem in preparing a cermet, and sometimes in joining larger ceramic and metal parts, is to get an adequate bond between a liquid metal and suitable ceramic. The ideal is to have adequate wetting, a slight chemical reaction and interdiffusion, surface keying and a favourable thermal expansion relationship on cooling. This problem occurs when single crystal ceramic whiskers are used to reinforce metallic alloys. Whiskers have been made in a long list of metals and ceramics and usually have strengths of several million psi, but to be effective the matrix in which they are embedded must have a strong bond to the whisker but it must not chemically attack it. Alumina whiskers have been shown to reinforce silver and a Ni-Cr alloy successfully to within a few hundred degrees of the melting point of the matrix metal, but little practical use has yet been made of whiskers.

Greater attention is at present focused on new high-strength polycrystalline ceramic fibres as a means of reinforcing materials, particularly plastics; continuous fibres of boron and carbon can be woven into cloth or even a three-dimensional body, before impregnation with a plastic matrix. These fibres have strengths of several hundred thousand psi and are of special interest because they are light and have a high Young's modulus of elasticity, which makes the composite very stiff. Glass-fibre reinforced plastics have a longer history, but the low elastic modulus of glass renders them very flexible so that a relatively large thickness and weight are required for a given degree of stiffness; this limits their attractiveness for aerospace applications.

Carbon fibre/silicon carbide composites have been used for rocket nozzles and other components requiring resistance to both thermal shock and high-temperature oxidation. The lubrication property of carbon, as graphite, has also been used by impregnating fibres to form composite materials for bearings. A less common type of composite is formed by using fine tungsten wires of 500 000 psi strength to reinforce TiC or oxide ceramics. The ceramic is thereby strengthened and is subsequently held together if cracking due to thermal shock or impact should occur.

The products of combustion expanding through a rocket nozzle and exhaust ducting may be essentially either reducing or oxidising; if they are the former, use can be made at 1500–2000° C of refractory metal alloys, carbides, borides and graphite. Some unusual composites have been made of graphite and refractory carbides which can be forged to shape at very high temperature and which have a remarkable combination of high strength and thermal shock resistance. In the case of oxidising gases at 2000° C, the only suitable materials are the refractory oxides, preferably present as a coating because of their poor thermal shock behaviour in thick sections. For nozzles a substantial layer of zirconia (melting point nearly 2700° C) has been coated onto tungsten alloys, the coating itself being reinforced with a mesh of Mo or W wires. For combustion chamber materials the temperature problem is less severe.

Modern methods of applying refractory coatings are 'flame spraying' and plasma-torch spraying. In the former, powder or a rod of the coating material is fed to an oxy-acetylene gun which projects the semi-molten or plastic particles at high velocity onto the prepared (microscopically rough) surface where they form a firm mechanical bond. This process is used to deposit layers of oxides, carbides and their cermets on various substrates which can be kept relatively cool. Best protection at the highest temperatures is probably obtained by depositing first an oxide cermet and then an outer layer of pure oxide. For nozzle applications, silicon carbide can be applied to graphite or tungsten by this process. Plasma spraying is a different technique which involves striking an electric arc of extremely high temperature in an inert gas jet and projecting the coating material in the liquid state in the inert gas 'plasma' (ion/electron mixture).

The high-temperature problems of re-entry vehicles apply to the nose cone and leading edges of the structure. The latter generate a high-temperature supersonic shock wave and may reach 1500° C or above for the short periods involved; here, suitably coated refractory alloys may be used to resist deformation and corrosion. For the nose cones of missiles, at which tremendous heat is generated, a technique is often used by which heat is disposed of by sacrificial volatilisation of an 'ablative material'. This should absorb the maximum heat energy possible before and during melting and finally in vaporising. Refractory oxides may be used, but certain plastics are good ablative materials and are sometimes combined or reinforced with ceramics. For instance, alumina or zirconia 'foams' of 80–90% porosity can be infiltrated with organic materials and used in this way. Such foams, by virtue of their high porosity, have good thermal shock resistance and are used also as heat shields.

11. Metals and Ceramics Tomorrow

Perhaps a little crystal gazing in metals and ceramics can tell us something of the future! A subject making some headway with a good deal of current research is superconductivity – the behaviour of certain metals and ceramics which lose all their electrical resistance at a critical temperature near absolute zero ($O° K$, or $-273° C$). An electric current continues to flow in a superconductor at very low temperature without dissipating any energy as heat; the prospect therefore arises of making electromagnets without a continuous power supply or of making cables or electrical machines in which there is no electrical loss due to resistance.

Several problems have so far restricted the practical applications of superconductivity; one difficulty is the very low temperature which must be employed, involving refrigerators using liquified gases such as helium, which boils at $4° K$. Another problem is the tendency of the magnetic field produced by the current to destroy the superconductivity effect. In all 'soft' superconductors, such as pure annealed niobium which has a relatively high critical temperature of $10° K$, the superconducting current is limited to a small value. A large current would produce a strong magnetic field which unfortunately would destroy the superconductivity; nevertheless, in niobium at $4° K$ useful direct and alternating currents are possible. Some 'hard' superconductors can withstand high magnetic fields and currents up to 100 000 amps/cm^2 at $4° K$ without loss of superconductivity. These materials contain crystalline defects or inhomogeneities which trap the magnetic field so that a large current can flow within the conductor, but they are essentially restricted to direct current use. Nb_3Sn and V_3Ga are 'hard' superconducting compounds and have a common crystal structure (beta tungsten); they are of special interest because their critical temperature at zero magnetic field is about 17° K, but unfortunately they are brittle and a long conductor offers fabrication problems. Other series of compounds of interest are carbides with the beta manganese structure, such as Mo_3Al_2C, and the niobium carbonitrides (NbC/NbN solid solutions); these have critical temperatures of about 10° and 17° K respectively. Such materials can sustain high currents at 4° K and it is possible that better compounds will be discovered. But at present the most developed hard (or 'high field') superconductors are ductile niobium alloys, particularly Nb-25–30% Zr and Nb-30% Ti, containing finely dispersed precipitates; again these have relatively high critical temperatures, but no superconductor has yet been found with a critical temperature above 20° K – the boiling point of liquid hydrogen.

At present, hard superconductor windings for large d.c. electromagnets are used for physics research and could be used for large d.c. electrical machines and for MHD generation of electricity. There is also a possibility that in future superconductors will be used in power transmission and transformers, but because of refrigeration costs it is likely that higher temperature superconductors must first be discovered.

Perhaps of equal interest at very low temperature are metals, such as aluminium and beryllium, which do not become superconducting, but in which electrical resistance is reduced by a factor of 10 000 or more, when they are extremely pure – say 99·999%. One could envisage aluminium a.c. conductors at 20° K giving very little electrical power loss as transformer windings, but it is again debatable whether the power saved is sufficient to offset the costs of refrigeration and the more sophisticated design.

Most pure metals are soft and ductile, but some nominally pure metals still have little ductility at room temperature. Beryllium in particular is difficult to prepare without some oxide and other minor impurites being present, and these impart a degree of brittleness. It is a light high-stiffness and high-strength metal of great potential as a structural material and as an electrical conductor; it also has possibilities of alloying to impart good high-temperature behaviour, for example in the aerospace field where it might compete with titanium. The metal at present has a few uses, but a highly alloyed beryllium (Be-40% Al) is available as a light ductile metal with the elastic stiffness of steel. The 'high temperature' BCC metals of Group VIB are equally notorious for their low-temperature brittleness unless very pure. Pure iron is ductile at much below room temperature, but there are still impurity and high-temperature gaseous contamination problems to overcome in Cr, Mo, W and their alloys and in Nb alloys. If we can be sure that thermal stresses or small impacts will not result in a metal cracking at room temperature, its potential use is very much greater.

The alkali BCC metals of Group IA have melting points little above room temperature, so they are normally rather soft and we do not encounter a brittleness problem. Sodium, despite its low melting point of 98° C, is of growing interest as an electrical conductor since it is relatively cheap and of very low density; a Na cable of equal electrical resistance to aluminium or copper is considerably lighter, but must be of greater diameter than either of these conventional conductors. The special problems of sodium cable development are its protection from atmospheric corrosion, adequate termination and joining of the cable, and the avoidance of faults which could produce failure by the overheating and melting of the sodium.

Current research is yielding a number of ultra-high-strength and quite tough steels which are certain to be utilised in structural applications. Certain carbon steels containing several percent of alloying additions can be strengthened considerably by mechanical working while in the austenitic range ('ausforming'), somewhat above the temperature at which they would transform from the FCC structure. Subsequent cooling or quenching and perhaps further mechanical working and tempering results in a single or mixed phase microstructure of small grain size, with fine carbide precipitates and tangled dislocations. Such steels can combine high hardness, yield points of about 250 000 psi, and moderate toughness; one possible application is as erosion-resistant parts of turbine blades which can be joined to the blade body by a neat electron-beam weld. Equally high strength with remarkable toughness is found in a class of very low-carbon, high-nickel content 'maraging' steels. A typical composition would be 18% Ni-8% Co-4% Mo-0·3% Ti. The steels are heat-treated, first by holding at about 800° C to dissolve the alloying elements, then by ageing at about 500° C, when the microstructure transforms from FCC austenite to a fine-grained intermediate crystal structure (martensite) which develops dislocations and large amounts of precipitate particles, notably of Ni_3Mo and Ni_3Ti, at the crystalline defects. Yield and tensile strengths of over 300 000 psi are possible, but presently developed steels which are also fairly ductile, tough and weldable have a yield and tensile strength in the 250 000 psi region. Maraging steels have very good fatigue resistance and are of interest for future aircraft applications. It has not yet been possible to make a stainless steel with these properties, for example by adding 10–15% chromium, and none of the ultra-high-strength steels retain their strength well at the elevated temperatures met in turbines and nuclear reactors.

High strength combined with very good high-temperature creep resistance is more likely to be found by further research and development on dispersion-strengthened metals. These are materials in which a fine dispersion of stable particles, such as oxides or borides, have the effect of blocking dislocation movement and so restricting slip and creep strain. The dislocation tangles which are formed also raise the recovery and recrystallisation temperature; theory suggests that the particles should be smaller than one micron in size to be effective. A few percent by weight of dispersant is often added as a separate constituent to the metal powder before it is recompacted and carefully worked, ideally so that the resultant material retains some ductility while its yield strength is increased. Alumina paticles have been used to dispersion strengthen aluminium and copper, while thoria (ThO_2) has been used with some success in nickel, cobalt and tungsten. A naturally occurring oxide on the particles of a metal or alloy powder may be sufficient to give useful dispersion strengthening on recompaction, for example with magnesium; sometimes it is possible to incorporate the dispersant directly into a liquid before casting or to use an electrodeposition or vapour deposition technique to get the correct combination of dispersant and metal.

Yield strength at room temperature can be increased considerably by dispersion strengthening, but high-temperature creep strength can be dramatically improved. Al_2O_3 particles dispersed in aluminium or copper raise the temperature at which the metal possesses a given creep strength by several hundred degress, while pure nickel containing dispersed thoria has creep properties at 800–900° C approaching those of the highly developed creep-resistant nickel alloys. Little research has yet been done with dispersants in creep-resistant alloys of nickel, iron or the refractory metals; research on such alloys to obtain higher temperature creep resistance without brittleness appears to be a field offering great potential – perhaps on nickel and niobium alloys for turbines, in particular.

Dispersion strengthening has been little used for nuclear reactor metals, although magnesium and aluminium have been considered in this respect. An obvious candidate for dispersion strengthening is zirconium, a relatively weak metal at high temperature, but needing great strength as an alloy for pressure tubes in water-cooled reactors. Since zirconium alloys are very reactive a dispersant of high chemical stability would be required – yttria (Y_2O_3) is a possible contender.

In developing a strong high-temperature alloy it is usually difficult to obtain a microstructure giving sufficient creep ductility. The problems may be due to minor impurities, grain size or the disposition of precipitates or dispersant in the microstructure resulting from the method of preparation used. Many of the principles involved to obtain the best microstructure are known, but the subject is so complex in its chemistry and physics that a clear solution to the problems is not always evident without a good deal of experimental work.

The preparation of large ingots of dispersion strengthened alloys which can be fabricated to form turbine rotors or tubes of suitable microstructure presents great problems. In large cast ingots of complex austenitic steels, segregation of alloying elements can occur which results in embrittlement and cracking when the ingots are forged; complex nickel alloys are equally difficult. Perhaps electroslag refining of steel and nickel alloys, where the ingot is built up by continuous addition and solidification, will provide a better prospect for large components of complex and dispersion strengthened alloys. But even if it is impossible to obtain large components of very complex alloys, it may be possible to join more simple shapes together by high quality very narrow welds which do not much affect the bulk properties of the composite component. Electron-beam welding can produce such narrow welds, many inches deep, between mating surfaces of alloys; plasma 'torch' and laser welding may also be developed in the future to join such thicknesses.

If the above developments occur, it is interesting to speculate on the future maximum temperature which would be acceptable in steam turbines and gas turbines. At present, steam turbines are limited to 565° C, partly because of boiler corrosion but mainly because the extra efficiency accruing from a reasonable increase in temperature would not compensate for the extra cost of stainless steels and complex alloys and processes involved. In the future it may be possible to make a very large increase in operating steam temperature, say to 700° or 750° C, so that the increase in turbine efficiency would more than pay for the better materials; if research and development prove successful, such materials should not be prohibitively expensive. Higher steam pressure and therefore higher working stresses, as well as temperatures, would apply to superheater tubes, steam pipes, valves, blades, rotor forgings, cylinders and bolts. Perhaps dispersion-strengthened complex nickel alloys combined with new methods of fabrication and joining could achieve this development. The two principal sources of heat for the 'boiler' – natural gas or the primary coolant of a high-temperature nuclear reactor –

should not involve the corrosion problem which is encountered with coal or oil.

Better materials and new fabrication methods should also lead to higher temperature and more efficient gas turbines, whether for electricity generation, aircraft or road vehicles. For electricity generation, the problem of turbine blade corrosion by ash from non-distillate oils would be avoided if use were made of natural gas or helium – the primary coolant of the HTGC nuclear reactor. For aircraft, protective silicide coatings on niobium alloy turbine blades may enable an advance to be made, while an alternative step would appear to be the reinforcement of complex nickel-chromium alloy blades with ceramic fibres or wires of higher-strength refractory alloys. Blades of silicon carbide or an oxide ceramic, perhaps based on alumina, are also possibilities, but the thermal shock problem of such non-ductile materials is known to present difficulties.

High-strength fibres and 'whiskers' of ceramics are likely to be combined with metal or ceramic matrices to provide a range of 'fibre composite' materials for aerospace and other specialist applications. Fibres of fused quartz, of several hundred thousand psi strength, have been used to demonstrate very effective reinforcement of aluminium when the fibre content is about 50%. Polycrystalline carbon fibres of equally high strength, but much greater elastic modulus, have been developed recently and have great potential for fibre composites which combine high strength, high stiffness and low density; carbon also has the unique advantage of retaining its strength up to 2500° C. An obvious problem in developing carbon fibre/metal composites will be the bonding and compatibility between the fibres and metals, particularly metals which readily form carbides; this may be solved by protecting the fibres with a stable coating, for example, of silicon carbide which might be bonded to the metal matrix by a thin SiO_2-metal oxide layer. Aluminium alloys and nickel alloys would appear to be the first candidates for carbon fibre reinforcement, since they could then be used for lightweight aircraft-skin structures and jet-engine blades respectively, both at higher temperature or stress. Similar reinforcement of steel, titanium and the refractory metals is possible, although these metals more readily form carbides at high temperature and the fibres would have to be well protected. Whiskers of silicon carbide, silicon nitride or alumina may also be used in future to form practical composites with metals; since the strength of whiskers is usually several million psi (ten times that of the special fibres), such composites should be even stronger, although perhaps expensive.

Suitable fibres or dispersed phases can be introduced into metals or ceramics to confer a variety of properties – increased

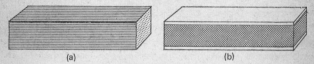

Fig. 27. *Materials of low density and high stiffness.* (a) *A low density plastic reinforced by high stiffness carbon or boron fibres.* (b) *A sandwich of low density foamed material (e.g. a plastic) between skins of high stiffness material (e.g. a metal).*

thermal conductivity, better lubrication behaviour, wear resistance, or special electrical or magnetic properties. In some instances, oriented needles of 'ceramic' precipitates, similar to single crystal whiskers, can be grown in an alloy by careful heat treatment. The most important new fibre composites, however, are likely to be those in which polycrystalline ceramic fibres confer high strength combined with low density and high stiffness. In this field plastics reinforced with high-strength carbon and boron fibres are already established as lightweight structural materials which are much stiffer than glass-fibre-reinforced plastics (Fig. 27). Greatest stiffness in a structural member for a given weight can be obtained in 'sandwich' composites which exploit the geometrical effect, for example, by employing a low-density porous core of some 'foamed' material sandwiched between strong skins of high elastic modulus, perhaps themselves fibre reinforced.

Current research on oxides and other ceramic compounds is

likely to result in better non-porous materials of controlled composition with a new range of properties. Full-density 'translucent' alumina is already important as a lamp envelope, but full-density beryllia (BeO) and magnesia (MgO) with melting points of 2550° and 2800° C respectively, have yet to be exploited. Both are good electrical insulators and are stable in an oxidising environment at 2000° C, when alumina is almost melting. The pure oxides have poor creep strength at this temperature, but it should be possible to strengthen them by fibres or dispersed phases. In particular, better magnesia bricks for steel furnaces, at 1600–1700° C, might be feasible if a suitable grade of magnesia could be reinforced with zirconia (ZrO_2) of melting point 2700° C. The detailed microstructure of refractory ceramics is in any case a subject of growing importance; means of controlling and producing new microstructures should result in refractory bricks with better resistance to corrosion and erosion.

Fully dense and transparent MgO has been produced by a hot-pressing technique in which a small amount of lithium fluoride (LiF) is added to the MgO powder. A few hundred degrees below its melting point MgO is sufficiently plastic to be forged to shape like a metal – an unusual operation for a ceramic. Polycrystalline magnesia is also unusual in that it has moderately good creep ductility above 1000° C; perhaps, despite its high thermal expansion, it will find new uses as a high-temperature transparent envelope or structural component. Yttria (Y_2O_3 – melting point 2410° C) containing 10% of thoria (melting point 3300° C) in solid solution has been recently sintered at 2000°–2200° C to form a fully dense transparent ceramic; zirconia and thoria are still the subject of much research and the whole field of oxides still offers considerable challenge.

Of the carbides, nitrides and borides, the borides are perhaps most intriguing; some are very hard and corrosion-resistant materials up to over 1200° C yet with very high electrical conductivity. Pure dense stoichiometric borides are difficult to fabricate, but when new methods of preparation do enable us to investigate borides more fully, some new applications are likely. Already, new boride cutting tools are being examined, and borides have been satisfactorily used as thermocouple sheaths in contact with liquid steel for many hours without excessive corrosion; their behaviour as stable high-temperature electrodes in corrosive media appears promising.

In the field of electronics, research on many new ferroelectric materials should be of interest, for example as piezoelectric transducers, or as transparent 'electro-optic' materials in which the applied voltage alters the character of the optical signal transmitted. As a transducer, potassium sodium niobate (PSN), a mixture of $KNbO_3$ and $NaNbO_3$, can be prepared as a fine-grained polycrystal which behaves efficiently at very high frequency; this may be of special interest in non-destructive ultrasonic testing, and in 'delay lines' in which electrical signals are temporarily stored as slower-moving acoustic signals. Single crystals of oxide compounds should lead to more efficient lasers and to their future use in communications systems, while new electro-optic materials (e.g. complex niobates) may be used in such systems for modulation. It is possible to prepare transparent polycrystalline glass-ceramics with interesting electro-optic properties by devitrifying a suitable glass containing a high percentage of ferroelectric oxide; a low nucleation temperature is used so that the glass crystallises with a very large number of small ferroelectric grains. Ferroelectric glass-ceramics can also be conveniently made possessing high permittivity; these may prove suitable for high-energy capacitors.

Some years ago it was found that electronic semiconduction could occur in certain glasses which contained ions of variable valency. For example borovanadate glasses, based on B_2O_3 and V_2O_5, contain ions of V^{4+} and V^{5+} which permit 'electron hopping' giving a moderate conductivity. This type of glass may be developable as a glaze for high-voltage insulator surfaces so that the voltage drop across them is uniform – a necessary condition to minimise unwanted electrical discharges when exposed insulators become dirty. Present 'semiconducting glazes' for outdoor insulators are a mixture of semiconducting crystalline oxides and insulating glass; these tend to corrode, notably by electrochemical effects, and there is need for a more stable material.

Very pure glass fibres have been considered for use as 'waveguides' for light from a laser which would be modulated to transmit data, speech or video signals; light waves are some 1000 times shorter than the microwaves used for beamed radar, telephone and TV transmission, and it would be possible to load a thin waveguide 'cable' with hundreds of thousands of simultaneous telephone conversations or many TV channels. For this application our best present optical glasses are of inadequate quality, and it would be necesary to prepare glass fibres of extreme purity and homogeneity with very low optical absorption. The feasibility of such futuristic optical fibre communications is still an open question, but much may depend on our ability to melt contamination-free glass and draw suitable fibres of it – perhaps composite fibres consisting of a one micron diameter inner core with a thicker annular coating of a glass of slightly different optical properties.

Although thin glass fibres are very strong, glass is normally limited by a very low strength in bending (about 10 000 psi) due to minute surface cracks which open up when a tensile stress is applied. In recent research new ways of strengthening glass are being devised which artificially introduce a high compressive stress in the glass surface, or prevent the cracks from growing by other means. One successful method for soda-containing glass is to exchange surface sodium ions for potassium ions from fused potassium nitrate (KNO_3) at 350° C; the potassium ions are larger than the sodium ions and so introduce a compressive stress in the surface which is largely retained on cooling the glass.

Another method applicable to sodium alumino-silicate glass is to carry out 'ion exchange' of sodium for lithium from a fused mixture of lithium and sodium sulphates (Li_2SO_4/Na_2SO_4). This produces a crystalline surface layer of beta eucryptite ($Li_2O.Al_2O_3.2SiO_2$) which has very low thermal expansion; on cooling, the glass below the surface of the body contracts much more than the surface layer, so that the latter is put in compression. Effective strengths up to 100 000 psi can be imparted by such processes, since the applied tensile or bending stress must first overcome the surface compressive stress before fracture. A different approach, which may render the material opaque, is to partially devitrify the glass at elevated temperature so that it contains dispersed small crystals which stop crack propagation, rather like dispersants in metals impede dislocation movement. Recent work has shown that 5000 Å size mullite crystals in a magnesium aluminosilicate glass raise its strength to nearly 100 000 psi (Fig. 28).

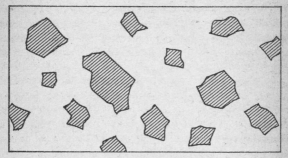

Fig. 28. *High strength part-devitrified glass. Precipitates of mullite crystals ($3Al_2O_3.2SiO_2$) in a magnesia/alumina/silica glass ($\times$ 20 000).*

High strength can also be obtained in fully devitrified glass-ceramics; a lithium silicate glass which devitrifies to give a strength of 30 000 psi has been developed, but other glass-ceramics giving much greater strengths should result from current research. Glass to metal seals can also be made with devitrifiable glasses, and these can subsequently be heat-treated to give strong ceramic-metal seals capable of withstanding a higher temperature than the parent glasses.

Glass-ceramics are quite versatile in that one can adjust the parent glass composition (including an additive to assist crystal nucleation) to give a range of tailor-made properties. Some glasses, containing ions of copper, silver or gold in trace quantities, are photo-sensitive in that only a pattern previously exposed to ultraviolet light will devitrify on heat treatment; this arises because the UV radiation produces local electron transfer converting the trace ions into atoms, which form precipitation nuclei for the ceramic. One lithium aluminosilicate glass-ceramic has almost zero thermal expansion between room temperature and 800° C; this material accordingly withstands severe thermal shock and has been demonstrated as a heat exchanger, initially fabricated in glass, for a gas turbine car. It is also possible to make certain fine-grain-size glass-ceramics which are transparent and therefore of possible future interest as high temperature lamp envelopes; present materials for such applications are fused quartz (usable up to about 900° C), or fully dense alumina which can withstand well over 1000° C. Perhaps a suitable transparent glass-ceramic could be found which is easier to prepare than either of these two materials.

A useful application of glass-ceramics which is likely to be applied in the building industry concerns the utilisation of waste slag from steel production. When sand is added to increase the slag's silica content, and a nucleating agent added, the resulting glass will devitrify to give strong wear-resistant tiles.

Further Reading

1. WULFF, J., *et al.* 1964–66. *The Structure and Properties of Materials.* Vol. I: *Structure.* Vol. II: *Thermodynamics of Structure.* Vol. III: *Mechanical Behaviour.* Vol. IV: *Electronic Properties.* Wiley, New York.
2. COTTRELL, A. H. 1964. *The Mechanical Properties of Matter.* Wiley, New York.
3. VAN VLACK, L. H. 1964. *Physical Ceramics for Engineers.* Addison-Wesley, New York.
4. WAYE, B. E. 1967. *Introduction to Technical Ceramics.* MacLaren, London.
5. GEMMILL, M. G. 1966. *Ferrous Alloys for High Temperature Use.* Newnes, London.
6. THOMPSON, G. V. E. and GATLAND, K. W. (eds.) 1963. *Materials in Space Technology.* Iliffe, London.
7. MACMILLAN, P. W. 1964. *Glass Ceramics.* Academic Press, New York.
8. 1965. *Reactor Materials. Proc. 3rd Int. Conf. peaceful Uses atom. Energy,* Vol. 9. United Nations.
9. 1965. *New Engineering Materials. Proc. Birmingham Conf.* Institution of Mechanical Engineers.
10. 1966. Metallic Materials. Symposium. *Jl R. aeronaut. Soc.* August.
11. ROTTENBURG, P. A. (ed.). 1966. *Advances in Materials. Proc. Instn chem. Engrs Symposium.* Pergamon Press, Oxford.
12. CARPENTER, L. G. 1964. Materials at high temperature. *Br. J. appl. Phys.* **15,** 871.
13. 1964. A discussion on new materials. *Proc. R. Soc.* A, **282,** 1–154.
14. 1967. Several articles on the nature and properties of metals, ceramics and other materials. *Scient. Am.* **217,** Sept.

Index

Ablative material, 115
Absolute zero of temperature, 29, 50
Acoustic waves, 75, 95, 125
Activation energy, 48
Ageing, 35–36, 67, 70
AGR reactor, 83–86
Alkali metals, 20, 99–100, 118
Alkaline earth metals, 20
Alloys, 25, 35–36, 41, 66–67
Alumina, 8–10, 38–39, 99–100, 113, 115, 119–120, 128
Aluminides, 106
Aluminium, 4, 39, 49, 79–80, 117, 124
Anion, 24, 94, 97
Arc melting, 42–43, 86
Ash, fuel oil, 66
Atmosphere, Earth's, 103
Atom, definition, 11
Ausforming, 118
Austenitic steels, 67, 86

Barium ferrite, 96
Barium titanate, 24, 93–95
Basic electric furnace, 2, 76
Battery, 97–98
BCC crystal, 22–23, 25, 30
Beryllia, 48, 100, 124
Beryllium, 85–86, 117–118
Beta alumina, 97
Beta eucryptite, 126
Beta manganese, 117
Beta tungsten, 117
Blades, turbine, 65–68, 73, 105, 122

Blanket, reactor, 87
Blast furnace, 2
Bolts, turbine, 72
Bonding, metal to ceramic, 99, 111, 113
Bonds, atomic, 17–21
Borides, 106, 111, 114, 124
Boron fibre, 113
  carbide, 109
  nitride, 110
Borosilicate glass, 31, 48
Borovanadate glass, 125
Brass, 3, 25
Brazing, 6
Brittleness, 11, 52–53, 68, 118

Cadmium sulphide, 93
Calcium tungstate, 100
Calder Hall, 60
Cans, fuel, 77, 79
Capacitor, 92, 95–96, 125
Carbides, 108–111
  (cemented), 61–64, 113–114
Carbon, 2–4, 18, 23, 46, 77, 109–110
  fibre, 113–114, 122–123
Castings, 4–6, 36, 75
Cast iron, 2
Cation, 23, 93–94, 97
Ceramics, traditional, 8–10
Cermets, 112–113
Chemical stability, 46
Chromium, 5, 46, 68, 71, 105, 118
Clay, 8–9, 43, 98
Coal-fired power stations, 60

INDEX

Coherent waves, 100
Cohesive strength, 47
Combustion chamber, 102, 104, 106
Compatibility, 79
Composite materials, 112–115, 122–123
Computer, 96
Contact materials, 49, 111
Control rods, 78
Copper, 3, 4, 28, 49
  in zirconium, 82
Cordierite, 99
Corrosion, 4–5, 46–47, 66, 73, 79, 102, 110
Corundum, 25
Cost of research, 58–59, 62, 64
Covalent bond, 17–21
CPH crystal, 22–23, 25
Cracks, 37–38, 75, 118, 126
Creep, 53–55, 66–68, 70–74
  ductility, 79, 85
Critical temperature of superconductor, 116–117
Crystals, 22–30
Cupro-nickel, 3, 5
Curie temperature, 94–95
Cutting tools, 61–64, 113, 125
Cylinders, turbine, 68–69, 72, 74
Czochralski technique, 91

Defects, 26–27
Deformation, 28
Delay lines, 125
Density, 45–46
Devitrified glass, 127–128
Diamond, 18, 23, 27, 110
Die-castings, 3
Dielectric, 95–96
Diffusion, 29, 35–36, 38, 48–49, 112
Diode, 89
Dipole, 94
Dislocation, 26–28, 36–38, 119
Dispersed particles, 82, 119, 123–124

Dispersion strengthening, 119–121
Domain, 51, 94, 96
Doped silicon, 89–91
Dragon reactor, 86

Efficiency, 69, 93
Elasticity, 46, 113
Elastic strain, 28–29, 37–38, 94
Electrical neutrality, 25, 27–28
Electrolysis, 97
Electrolyte, 97
Electrolytic deposition, 3–5
Electron, 11–21, 88–89, 97
  hopping, 93–94, 125
Electron beam welding, 7, 119, 121
Electron microscope, 39–40
Electron probe microanalyser, 41, 81
Electron spin, 50, 96
Electronegativity, 16, 47
Electronics, 88–96, 125
Electro-optic material, 125
Electrophoresis, 43
Electroslag refining, 7, 121
Electroslag welding, 7, 76
Elements, 1, 11–17
Enamelling, 9
Energy, 12–13, 97
Epitaxial deposition, 91
Erosion, water droplet, 73, 119
Etching, 33
Extrinsic semiconductor, 88

Fast reactor, 61, 86–87
Fatigue, 55–56
FCC crystal, 22–25, 30
Feldspar, 8, 98
Ferrimagnetic ceramics, 96
Ferrites, 96
Ferroelectric ceramics, 94–96, 125
Fibres, 113–114, 122–123, 126
Field ion microscope, 40
Fission, nuclear, 77–79

Flame spraying, 114
Floating zone refining, 90–91
Fluorite, 24–25, 97
Foams, 115
Forged material, 5, 36
Forsterite, 99
Fracture strength, 28, 36, 38, 51
Free-electron bond, 17–21
Fuel cell, 97–98
Fuel element, nuclear, 77
Full density ceramics, 38, 99–100, 124, 128

Gallium arsenide, 88, 92
Galvanising, 3
Gamma-ray photographs, 75
Gas-cooled reactor, 77–86
Gas-pressure bonding, 111
Gas turbine, 65–68, 122
  car, 128
Generation cost, electrical, 60–61
Germanium, 88
Glass, 8–9, 30–32, 38–39, 55, 125–128
Glass-ceramic, 99, 125–128
Glass-fibre plastics, 113
Glaze, 9, 125
Government research, 58–59
Grain boundary, 26–28, 33, 36–38
  growth, 36–38, 100, 112
  size, 36–38, 99
Graphite, 86, 109–110, 115
Graphitisation, 71
Gravitational field, Earth's, 102–103
Griffith cracks, 38
Growth (of uranium), 79

Haffnium, 104
Hair-line cracks, 76
Halogen elements, 108
Hard ferrites, 96
Hard superconductors, 116–117

Hardening of rotor, 70, 74
Heat treatment, 67, 70
Heavy water, 82
High rate forming, 6
High-strain fatigue, 55, 75
High temperature metallurgy, 65–68
Hot pressing, 111–112, 124
HTGC reactor, 86
Hydrogen
  in magnesium, 82
  in steel, 75–76
  in zirconium, 83
Hydrostatic extrusion, 6

Induction heating, 42, 90–91
Industrial research, 58–59
Inorganic materials, 1, 20
Insulator, 31, 98–99, 125
Intermetallic compound, 18, 25
Interstitial atom, 25–26
Intrinsic properties of materials, 45
Intrinsic semiconductor, 88
Iodine, 20–21, 107–108
Ion, 16–19, 97–98, 126–128
  exchange, 126
Ionic bond, 17–19, 21
Iron, 2–3, 30, 118
Isostatic hot pressing, 111

Jet engine, 65–66, 122
Joining, 6, 99, 113

Kaolinite, 98
Kinetic energy, 29–30

Lamp envelope, 128
Lamp filament, 105, 107–108
Large-scale research, 58–61
Laser, 7, 41, 100–101, 125–126
Lattice, crystal, 25–26
Lead, 3
  glass, 32
Light meter, 93

INDEX

Lithium, 46
  in glass, 126–127
  in MnO, 93
Loss, (di-)electric, 96, 99–100
Low-strain fatigue, 56

Machining, 6, 61–63
Magnesia, 2, 9, 53, 124
Magnesium, 4, 80–82
Magnesium-manganese ferrite, 96
Magnetic field (and superconductors), 116–117
Magnetic materials, 50, 96
Magnetite, 93
Magnox AL80, 80
Manganese (in magnesium), 80–81
Manganese oxide, 93
Maraging steel, 119
Mass spectrometer, 40–41
Materials science, 57–58
Matrix, lattice, 33, 35
Mechanical behaviour, 50–56
Mechanical treatment, 5
Melting point, 47
Metalliding, 106
Meteorite, 35, 103
MHD generation (of electricity), 100, 111, 117
Microelectronics, 92
Microstructure, 33–44
Microwaves, 96, 100–101, 110, 126
Miniaturisation, 92
Moderator, 78, 110
Modulated waves, 101, 125–126
Molybdenum, 105–107, 114, 118
  in steel, 71–74
  in zirconium, 82
Mond process, 5
Mullite, 98–99

Natural gas, 61, 98
Neodymium, 100

Neutron, 77–79
  absorption, 78–79
Nickel, 3–5, 39, 67, 113
  in steel, 70–71
Nickel silver, 3
Nimonic alloys, 67
Niobates, 125
Niobium, 105–106, 116, 118
Nitrides, 110–111
Non-destructive testing, 75, 95
Nose cone, 103, 115
Nozzle (rocket), 102, 104, 114–115
n-type semiconductor, 89–92
Nuclear reactor (or pile), 59–61, 77–87, 100, 109, 121–122
Nucleation, 27, 30–31, 34–35, 99

Optical microscopy, 33, 39
Organic solids, 18

Periodic table, 12–16, 20
Permittivity, 95
Perovskite, 24
Phase, 25, 33–35, 38–39
  diagram, 34, 38, 81
Photoconductive properties, 93
Photosensitive glass, 127–128
Piezoelectric ceramics, 94–95, 125
Plasma torch, 114–115
Plasticity, 27, 36–37
Platinum, 17, 104
Plutonium, 15, 60, 80, 87, 109
p–n junction, 89–93
Polarisation, 94–96
Poling, 94
Polycrystal, 26–28, 37–38, 40–41
Polygonisation, 36
Porcelain, 8–9, 39, 98–99
Porosity, 38, 43, 75, 99
Portland cement, 9
Powder metallurgy, 7–8

Precipitation, hardening or strengthening, 33–35, 66–67, 72
in glass, 127–128
Pressure tube, 82–83, 120
Pressure vessel, 77, 82
Protective coating, 103, 105–107, 110, 114
p-type semiconductor, 89–92
Pyrolytic deposition, 44, 86, 109–110

Quantum mechanical laws, 13
Quartz, 8, 31, 98, 122, 128
Quartz-iodine lamp, 108
Quenching, 35, 70–71, 79

Radar, 96, 100
Rare-earth elements, 13–15
Recovery, 36–37, 54, 119
Recrystallisation, 36, 67, 119
Rectifier, 89
Re-entry (atmospheric), 103–104, 115
Refractories, 2, 10
Refractory metals, 102, 104–108, 114–115
Resistance (electrical), 49
Reverse voltage (of diode), 89–90
Rhenium, 107
Rocket, 102–104
Rocksalt, 24, 27–28
Rolling, 5
Rotor, turbine, 65–70, 73–76
Ruby, 100

Sand, 8, 98
Sandwich materials, 123
Sapphire, 27
Satellite, 102–103
Seals, 127
Seed crystal, 91
Self diffusion, 48–49
Semiconductor, 88–94, 125–126
SGHW reactor, 82
Shell, electron, 13–18

Silica, 8–10, 31
Silicides, 105–106
Silicon, 88–92
carbide, 10, 86, 109, 111, 114–115, 123
nitride, 110, 123
Single crystal, 27–28, 33, 37, 41, 68, 90–91, 100, 125
Sintering, 7, 9, 38, 43
Slip, 27–28, 37
Slip casting, 43
Small-scale research, 58–59, 61–64
Soda-lime glass, 8, 31, 48, 98
Sodium, 49, 118
in battery, 97
in fast reactor, 87
in glass, 126
Soft ferrites, 96
Soft superconductors, 116
Solar battery, 93
Solder, 3, 6
Solid solution, 25, 33, 35, 37
Space technology, 102–115
Space vehicle, 102–104
Sparking plug, 98, 111
Spinel, 24, 96
Spray refining, 2–3
Stainless steel, 3, 33
Steam turbine, 68–76
Steatite, 99
Steel, 2–3, 6, 25, 29, 39, 46, 67, 70–76, 86–87, 119
Stiffness, 46, 113–114, 118
Stoichiometric compound, 27
Strain energy, 26, 28, 36
Strain, plastic, 28, 36–37
Strength, 28, 36, 38–39
Stress, 28–29, 38
Stress–strain curve, 51, 53
Substrate, 92, 100, 109, 114
Superconductivity, 50, 116–117
Superheater (tubes), 68–70
Superplasticity, 55
Surface energy, 36, 47
Swelling (of uranium), 78–79

INDEX

Tantalum, 105–106
  carbide, 62, 109
Tellurides, 92
Temperature, 29
Tempering, 70–71, 74
Tensile strength, 28, 39
Thermal
  conductivity, 29–30, 32, 47–48, 109
  expansion, 47–48, 110, 126, 128
  shock, 10, 48, 102, 104, 106, 109–110, 114–115, 128
  stress, 118
  treatment, 5
  vibration (agitation), 29–30, 53–54
Thermionic generator, 100
Thermistor, 93
Thermocouple, 104, 107
Thermodynamics, 30, 34
Thermoelectric properties, 92
Thoria, 10, 98, 119–120, 124
Tiles, 9, 128
Tin, 3
Titanium, 46, 104–105
  carbide, 62, 109, 112–114
  diboride, 111
Toughness, 52
Transducer, 94, 125
Transistor, 90
Transition temperature, 52–53, 68, 74, 105
Transparent ceramics, 100, 124–125, 128
Tungsten, 105–108, 114–115, 118
  carbide, 62, 109
Tungsten halogen lamp, 108
Turbine power plant, 58, 61, 65–76, 100, 120–122

U.K. Atomic Energy Authority, 60
Ultrasonic testing, 75–76
  vibrations, 6, 75

Ultraviolet (UV) light, 128
University research, 58
Uranium, 77–81
  carbide, 109
  dioxide (urania), 25, 27, 82, 84–87, 109

Vacancy, 26–27
Valence electron, 15
Van der Waals bond, 20
Vanadium, 105
  in steel, 71–73
  pentoxide, 66

Waveguide, 126
Wear resistance, 62–63, 109, 111, 113
Welding, 6–7, 70, 103
Wheels, turbine, 74
Whiskers, 37–38, 113, 122–123
Widmanstätten structure, 35
Work hardening, 28, 36–37, 54, 67
Wrought product, 4–5
Wurtzite, 25

X-ray diffraction, 40
  photographs, 75
X-rays (from Sun), 103

Yield strength (or stress), 28, 37, 51
Young's modulus of elasticity, 29, 46, 113
Yttria, 84–85, 120, 124

ZA alloy, 80–82
Zinc, 3
Zircaloy, 83
Zircon, 99
Zirconia, 9–10, 98, 114, 124
  electrolyte, 97
Zirconium, 46–47, 82–83, 104, 115, 120
  diboride, 111
  in magnesium, 80–82
Zone refining, 43, 90–91